Integrated High Resolution Imaging Radar and Decision Support System for the Rehabilitation of WATER PIPElines

Integrated High Resolution Imaging Radar and Decision Support System for the Rehabilitation of WATER PIPElines

Edited by
Matthaios Bimpas, Angelos Amditis and Nikolaos Uzunoglu

Publishing
London • New York

Published by IWA Publishing
Alliance House
12 Caxton Street
London SW1H 0QS, UK
Telephone: +44 (0)20 7654 5500
Fax: +44 (0)20 7654 5555
Email: publications@iwap.co.uk
Web: www.iwapublishing.com

First published 2010

Typeset in India by OKS Prepress Services.

British Library Cataloguing in Publication Data
A CIP catalogue record for this book is available from the British Library

Library of Congress Cataloging-in-Publication Data
A catalog record for this book is available from the Library of Congress

ISBN 10: 1843393719
ISBN 13: 9781843393719

Contents

 Integrated High Resolution imaging Radar and Decision Support System for the rehabilitation of WATER PIPElines. Edited by Matthaios Bimpas, Angelos Amditis and Nikolaos Uzunoglu. ISBN: 978184333719. Published by IWA Publishing, London, UK.

Abbreviation list

A/D-C	Analogue to Digital Converter
AC	Asbestos Cement
ADO	ActiveX Data Object
AGAP	Association pour la qualité en Géophysique Appliquée non Pétrolière
ALC	Active Leakage Control
API	Application Programming Interface
ASCE	American Society of Civil Engineers
ASTM	American Society of Testing and Materials
AWWA	American Water Works Association
AWWARF	American Water Works Association Research Foundation
AZNP	Average Zone Night Pressure
CARL	Current Annual Real Losses
CEI	Comitato Elettotecnico Italiano
CMMS	Computerized Maintenance Management System
CW	Continuous Wave
DB	Database
DAO	Data Access Object
DM	Data Manager
DMA	District Metered Area
DOE	Department of the Environment (UK)
DSS	Decision Support System
DWI	Drinking Water Inspectorate (UK)
ELL	Economic Level of Leakage
E/M	Electromagnetic
EPA	Environment Protection Agency
ES	Expert System
ESA	European Space Agency
ETSI	European Telecommunications Standards Institute
EU	European
FFT	Fast Fourier Transform
FP	Framework Programme (European Union)
GIS	Geographic Information System
GPIR	Ground Probing Imaging Radar

GPR	Ground Penetrating Radar
GSM	Global System for Mobile Communications (cellular phone technology)
GUI	Graphical User Interface
HDD	Horizontal Directional Drilling
HDPE	High Density Polyethylene Thermoplastic Pipe
IT	Information Technology
IWA	International Water Association
KB	Knowledge Base
KPI	Key Performance Indicator
LFR	Leakage Flow Rate
LNA	Low Noise Amplifier
MAS	Method of Auxiliary Sources
MDPE	Medium Density Polyethylene
MNF	Minimum Night Flow
NDF	Night-Day Factor
NPV	Net Present Value
NRW	Non-Revenue Water
O&M	Operations and Maintenance
PA	Power Amplifier
PC	Personal Computer
PCB	Printed Circuit Board
PDA	Personal Digital Assistant
PE	Polyéthylène
PRF	Pulse Repetition Frequency
PRV	Pressure Regulating Valve
PSTN	Public Switched Telecommunications Network
PVC	Polyvinyl Chloride
RPP	Reinforced Polyester Pipe
SMS	Short Message Service
SNR	Signal to Noise Ratio
TEM	Transverse Electromagnetic
TSA	The Survey Association
TTL	Transistor-Transistor Logic
UARL	Unavoidable Real Losses
UFW	Unaccounted-For Water
UI	User Interface
ULF	Ultrac Leak Finder
UWB	Ultra Wide Band
VCO	Voltage Controlled Oscillator
VTK	Visualization ToolKit
WIS	Water Into Supply
WRC	Water Research Center
WSSTP	Water Supply and Sanitation Technology Platform

Editorial

Matthaios Bimpas, Angelos Amditis and Nikolaos Uzunoglu

> *"The earth counts on the water and everything is water..."*
>
> *Thales of Miletus*

Many European (EU) cities are experiencing increasing problems with their water pipeline infrastructure. The cost of replacing these old, worn-out systems, if left to deteriorate beyond repair, is astronomical and clearly beyond the resources of many communities. Replacement, however, is not the only choice, as many of these systems can be rehabilitated at 30 up to 70 percent (30–70%) of the cost of replacement. Accordingly, resources allocated to address pipeline rehabilitation management issues are increasing. Due to the emphasis on sustainable management, risk-based approaches for the rehabilitation management of the water supply network need to be developed, as well as technologies for the inspection and evaluation of the pipeline conditions.

This has been the overall aim of the WATERPIPE project

WATERPIPE project has contributed to the topic II.3.3 "Advanced Technologies for Locating, Maintaining and Rehabilitating Buried Infrastructures" of the "Global Change and Ecosystems" thematic sub-priority of the FP6, by developing new, reliable technologies for water distribution -incorporating also performance and risk based approaches- for locating assets, identifying defects and leaks, monitoring and rehabilitating the buried infrastructures.

The main aim was to improve operation, rehabilitation, serviceability, pollution prevention and safety and thus minimising direct and indirect costs, including the environmental and socio-economic ones

WATERPIPE equipment were designed to provide a three dimensional representation of the underground pipeline network and at the same time provide information about the condition of the network; in this way, WATERPIPE system initially aimed to detect leaks that are not visible and are applicable to both metallic and non-metallic pipelines. The system has been developed to work synergically with other devices in order to achieve better, more accurate and more efficient results. Its use aimed to reduce water loss through leakage control and promote a water efficient, environmentally sustainable water industry also by mitigating the impacts of global change.

The high cost of developing additional water resources means that other ways, including leakage reduction, must be explored within existing resources. The project results consist in water savings through leakage control and thus promoting economic efficiency and reductions in energy use.

Research undertaken within WATERPIPE intended to strengthen the scientific and technological capacities needed in Europe to control the impact of global change on water resources availability. The project has developed devices, methodologies and decision-support tools for achieving an integrated management of water supply systems taking into consideration socio-economic and technological aspects of water use.

WATERPIPE has promoted a pro-active approach in the management of the underground pipelines of water utilities that conserves water and energy. Moreover, by extending the life of the pipelines it is decreasing the use of raw materials and increasing their conservation.

Last but not least, WATERPIPE has contributed to establish Europe as the international scientific leader on issuess of sustainable development in respect to the water cycle.

Chapter 1

The problem and the main objectives of WATERPIPE project

Matthaios Bimpas, Mata Frondistou, Nicola Bazzuro, Angelos Amditis, Alan Jones and Nikolaos Uzunoglu

THE PROBLEM

Water demand is increasing every year by two up to three percent (2–3%), while at the same time ***global warming*** may possibly lead to a significant shrinkage of the groundwater content available water. At the same time a considerable amount of the water provided by EU companies is lost through leakage. Water lost through leakage means higher abstractions and an even greater strain on the water environment.

More than 32 billion cubic meters (m^3) of treated water is lost through leakage[1] from water distribution networks each year. One estimate places the worldwide cost to the water utility companies for that leakage at 6 billion US dollars[2].

In the United Kingdom (UK), Thames water which supplies the majority of potable water for London spends an average of £500,000 a day fighting leakage. Much of this goes on replacing around 500 km of corroded cast iron water mains each year.

In California, United States of America (USA); 18% of the energy used by the entire state is consumed in the provision of its water supply[3].

In some low-income countries, this loss represents 50–60% of water supplied, with a global average estimated at 35%. Saving just half of this amount would enable the supply of water to an additional 90 million people without any further investment.

At Aguas do Porto, Portugal, the non revenue water volume was halved in 8 months, almost entirely due to the reduction of leakage[4].

Having a water leak reduction strategy is clearly a must for any provider of potable water, whether they are public or private. However, the cost of implementing that strategy is a significant factor in its long term success.

---

[1] The loss of water from a distribution network between the point of water treatment and the point of delivery to the consumer.

[2] Extrapolated from the World Bank Discussion Paper no 8 – December 2006.

[3] International Water Association (IWA) Task Force on Water Utility Efficiency in Low and Middle Income Countries (LAMIC TF) arbon Credit Concept Note v6.

[4] 2008 Global Leakage Summit.

 Integrated High Resolution imaging Radar and Decision Support System for the rehabilitation of WATER PIPElines. Edited by Matthaios Bimpas, Angelos Amditis and Nikolaos Uzunoglu. ISBN: 978184333719. Published by IWA Publishing, London, UK.

Several water utilities do not make any attempt to locate leaks other than those that appear on the surface and are easily visible to staff or members of the public. Other water utilities use, in the overwhelming majority of cases, acoustic equipment to locate leaks. Although acoustic equipment is generally considered satisfactory for metallic pipes, its application to plastic and asbestos cement pipes is often problematic.

WATERPIPE promotes a pro-active approach in the management of the underground pipelines of water utilities that conserves water and energy. Compared to the Ground Penetrating Radar (GPR), which is a non-intrusive investigation tool used for locating buried utility infrastructure, the WATERPIPE Ground Penetrating Imaging Radar (GPIR) systems take the processing of data one stage further by using modern 3-Dimensional (3-D) imaging techniques. WATERPIPE also includes a Decision Support System for the rehabilitation of water pipelines.

As far as the Decision-Support-System (DSS) is concerned, the output of inspection equipment that detects leaks and/or cracks can be used as input to the DSS that makes use of them in order to:

- Assess probabilistically the structural integrity and reliability of the pipe as a function of time (through the Structural Reliability Module).
- Assess probabilistically the leakage rate as a function of time (through the Leakage Module).
- Assess probabilistically the level of water contamination and compliance with regulation (through the Water Contamination Module).

The DSS additionally contains an Expert System (ES) that acts as an intelligent intermediary between the user and the results that can be obtained by the system.

OVERVIEW OF THE BUSINESS OPPORTUNITY

Water leakage costs

The non revenue water losses within European water companies run into millions of Euros every year and World wide, they run into billions of dollars. It is no surprise therefore that with the cost of supplying potable water set to rise year on year; the reduction of non revenue water loss is seen as a top priority around the world. Figures 1.1 and 1.2 below, give some indication of the scale of the problem.

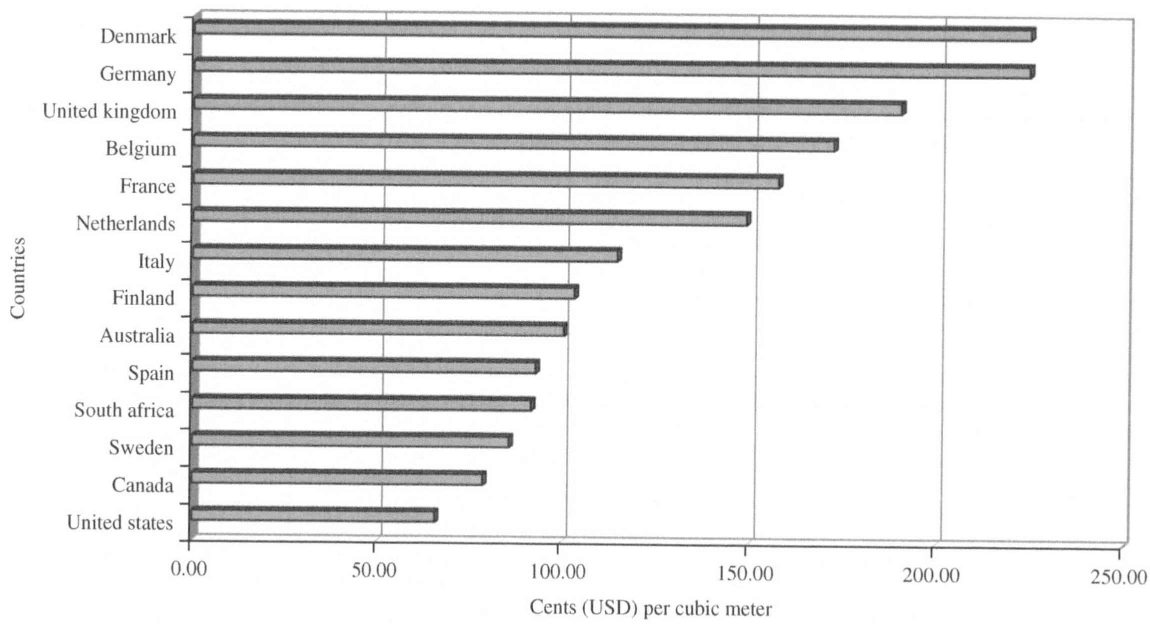

Figure 1.1 International water cost comparisons (2006).

It is not however the repairing of the actual leak which normally present the problem. It is the locating of that leak in the first place. No easy task when faced with a supply network stretching hundreds of kilometres (kms). Once located, the next problem is quantifying the leak and assessing the priority for repair. This includes such factors as the significance of its position, (both topographically and with respect to the supply network); the overall cost of the lost water against the overall cost of the repair and the timeframe involved.

City or country	Current NRW %	Targets mentioned
Malaysia	36.4	NRW 25% in 2005
Italy	30 to 40	'Acceptable' Long-term NRW 15%
Korea	29.3	NRW 15% in 2001
Japan		Real Losses 5% to 10%
Finland	12 to 25	NRW under 10%
France rural France urban	10 to 40 10 to 30	Lowest compatible with economic balance
Poland	<10 to >20	System-specific based on consumption (Fig 3)
Romania		Real Lossed 15% to 20%, next 10 to 20 years
Hong Kong	30.5	Real Lossed 16% in 20 years time
Portugal	18 to 58	Not stated
Murcia, Spain	9.7	Not stated
Norway	40	NRW less than 20% would not be economic
Denmark	7.6	NRW <10%, tax of 5 Kr/m3 if above this
Bangkok, Thailand	38.8	NRW not to exceed 30% in 2004
Hungary	24	Not stated
USA	5 to 37	Target systems with NRW > 10%
Taiwan Province	23.1	Not stated
Taipei, Taiwan	41.8	Not stated

Figure 1.2 References from National reports to % of non revenue water losses.

Current methods of detection

Although several methods of leak detection are now available and in daily use around the world, none provides an ideal solution for all networks. An analytical description of the current methods is provided in the following chapters; however a brief description of these methods is given below:

Acoustic systems

Acoustic systems for example work vey well on cast iron pipe networks, but are less effective on more modern networks made from materials like Medium Density Polyethylene (MDPE).

Flow monitoring

Flow monitoring particularly at night time or periods of low use can highlight areas of concern, but its use is limited to surface access points, therefore restricting the pinpointing of any leak to at best several metres of pipe, and in worst case scenarios, many hundreds of metres.

Common drawback

All current methods have one common drawback. They only work after the fact i.e., "The leak must exist before it can be detected and it must be a big enough leak to be detectable". This means that in the majority of cases a substantial loss of water from the system has to occur before it can be detected and repaired. In many cases this has meant small water losses remaining unnoticed for years, resulting in an overall significant loss to the system.

Additional problems

Damage to water infrastructure

In addition to the significant losses of potable water, damage caused by the lost water is often, also significant; with pipes being undermined along several metres reducing the lifetime of that pipe and the resulting lack of support causing additional leaks to appear at each joint; amplifying the water loss even further and increasing rehabilitation costs.

Damage to adjacent infrastructure

Damage to adjacent services such as communications networks and electrical supplies is quite common, as is damage to nearby properties. Large sections of the highway above such leaks have been substantially affected and in extreme cases, collapsed entirely.

Drivers for change

Under the European (EU) Water Framework Directive (2000) and supported by a raft of subsequent National and European legislation; the control and management of water quantity is now a legal requirement across Europe. For any company or public body responsible for achieving this objective, an essential pre-requisite to success is improving water use efficiency. One major element of which, must be the management and reduction of non revenue water loss, from their water supply system.

Within the water industry many ways of managing water loss exist. For the purposes of this simple overview however they may all be regarded as falling into one or more of the following groups[5].

(1) Part of an active Pipeline & Assets Management program. i.e., planned maintenance and rehabilitation.
(2) Part of a Pressure Management policy. Often proven to extend the useful life of ageing network infrastructure, reduce the frequency of new leaks occurring and the rate of loss once occurred.
(3) The speed and quality control of any repair work undertaken i.e., the time taken from a leak being reported, to the completion of an effective repair or replacement and normal service restored.
(4) Part of an active leakage control programme, i.e., night time flow monitoring etc.

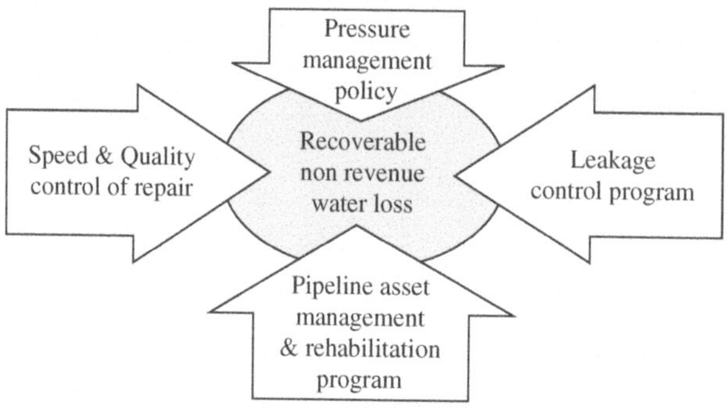

The WATERPIPE objectives

The overall objectives of the WATERPIPE project were:

- **To develop a novel, high resolution imaging ground penetrating radar** for the detection of pipes, leaks and damages, for the identification of the damaged region and the evaluation of it at a test site.
- **To produce an integrated system** that contains the equipment above and a DSS for the rehabilitation management of the underground water pipelines that uses input from the inspections to assess, probabilistically, the time dependent leakage and structural reliability of the pipelines and a risk-based methodology for rehabilitation decisions that considers the overall risk, including financial, social and environmental criteria.
- To field test the equipment and the DSS.

To accomplish the above, specific objectives of WATERPIPE project were set:

(1) To develop a high resolution Ground Penetrating Imaging Radar (GPIR) for detecting water pipes, including their size and material, for detecting leaks and damages in such pipes, for imaging the damaged region and for evaluating the performance of the above GPIR in a laboratory.
   - The above GPIR is able to detect water pipes made of all types of materials. It is also able to detect leaks and damages in water pipelines of all types of materials. Due to the fact that this was the main technological challenge of the system, the technological partners of WATERPIPE suggested the use of additional devices to detect and locate the water leaks, in case of damages in the pipelines.
   - The penetration capability of the system is up to 2 meters into the ground.
   - The image resolution capabilities of the damaged pipe is less than 5 cm.
   - The survey time (detection and imaging) is approximately 10 sec/meter along the pipe axis. It takes almost 3 hours to survey a kilometre of the pipeline.
   - The GPIR is mountable on a 4×4 wheel car. A laptop computer is also inside the car while the antenna system and the transceiver-analogue/digital signal processing units are placed in the front car.
   - The antenna is of a square shape of 1×1 m in horizontal dimensions while the thickness is about 10 cm.
   - In each position, the system is able to survey a length of 5 meters along the pipeline. Thus, to survey a pipe length of 100 meters, the car must stop at 100 divided by 5 equals to 20 points along the pipeline.

[5] International Report on Water Loss Management and Techniques, Allan Lambert, IWDC Ltd, 12/04/2004.

(2) To develop a methodology to use the results of inspection with the GPIR, in order to assess the flow rate of leakage at the time of the inspection and to predict, probabilistically, the flow rate of leakage in future operating periods taking into account the growth of the size of defects. Such growth is based on the accumulation of leakage and the structural condition of the pipe. This prediction is updated when new data from inspections on the size of the defects become available using a Bayesian updating procedure.

(3) To develop a methodology to assess probabilistically and as a function of time the concentrations of pathogens and other contaminants due to loss of physical integrity of the pipeline based on the growth of the size of the derived defects and the location of the pipeline.

(4) To develop a methodology to use the results of inspection with the GPIR (horizontal and vertical cracks and deformation) to assess the structural reliability of the pipeline at the time of the inspection and predict, probabilistically, the structural reliability of the pipeline during future operating periods taking into account the development of loads and the growth of deterioration by the various degradation mechanisms and uncertainties. Forces on a cross-section vertical to the pipe axis are being analysed, in order to assess the chance of crushing failure (bursts). Forces along the axis of the pipe are also analysed to assess the probability of bending and joint failure for each pipe material. Structural limit states are determined in terms of allowable stresses and strains. Probability of structural failure is taken to mean probability of exceeding the above limits.

(5) To develop a risk based methodology for the rehabilitation management of the trunk and distribution water mains. This methodology attempts to answer the following questions:
   - Which of the leaking and/or structurally deteriorating pipelines should be rehabilitated?
   - When and how it will be done, for each one of them?
   - When should individual pipelines be re-inspected?

(6) To produce an integrated package that contains the GPIR and a DSS for the rehabilitation management of water pipelines. The output of the GPIR feeds automatically into the DSS for an assessment of leakage, contamination and structural adequacy and based on these to decide on the method and timing of rehabilitation. The DSS includes:
   - a software implementation that determines, probabilistically, the flow rate of pipe and joint leaks over time (point 2 above);
   - a software implementation that provides a probabilistic, time dependent assessment of the concentrations of pathogens and other contaminants because of loss of physical integrity of the pipeline (point 3 above);
   - a software implementation that determines, probabilistically, the time-dependent structural reliability (and thus, also, probability of failure) of pipes and joints (point 4 above);
   - a software implementation that allows water utility managers to decide which of the pipelines to rehabilitate, when and how and when to re-inspect each pipeline based on the expected risk levels (point 5 above);
   - expert system technology for the integration of the different modules and a flexible User Interface (UI);
   - a Geographic Information System (GIS) module to examine specific areas of pipes and evaluate the results in a visual manner taking into account the effects of pipe location, soil type, etc.
   - a database that serves as the repository of all protected related data;
   - a data manager which is the only access point to the database and which provides a customised view for each particular module with access control mechanisms which assure the data integrity in the database.

(7) To test in the field the integrated package above on the networks of selected water utilities. In addition, the GPIR is field tested in blind tests and evaluation of the DSS have been carried out for ease of use, user-friendliness, relevance, completeness, and cost effectiveness.

DESCRIPTION OF THE BOOK STRUCTURE

This book has been grouped into chapters, based on the kind of topic covered and its nature (e.g., introduction, description, use or case study).

Depending on the topic to be covered, the chapter can be addressed to different level of professionals involved in the execution of leakage detection (managers, technicians, etc.) and the characteristics and level of development of the water distribution network. The chapters incorporate the information required for acquiring knowledge on asset management, leakage detection and quantification as well as on collecting data with the GPIR and process them through the DSS.

Each chapter represents a module that is self-contained, written such that it can be directed to different types of target professionals in the field, according to their needs. The modules contain as much information as is required to achieve their particular objectives, and to provide a high degree of flexibility when assembled for use in example for training courses.

This book begins by providing the water practitioner with general information on asset management. It examines the assessment and techniques required for leakage quantification. It then goes on to show the various equipments, techniques, operating procedures, methodologies etc. as well as some case study applications.

The content of each chapter can be summarized as follows:

- Chapter 1: The problem and the main objectives of WATERPIPE project
- Chapter 2: Introduction to the asset management
- Chapter 3: Leakage monitoring and control: pressure management, detection and location of leaks
- Chapter 4: Equipment for leak detection and location
- Chapter 5: Usage of GPIR devices: the WATERPIPE ground penetrating radar
- Chapter 6: Measurements with the WATERPIPE GPIR system and the LEAKING system
- Chapter 7: Design, implementation, testing and refinement of the expert system, the data base, the data manager and the geographic information system GIS module
- Chapter 8: Testing Report for the Decision Support System

The present manual gives a practical demonstration of the available equipment (GPIR and DSS) over a range of technology, followed by field demonstration.

REFERENCES

Malcom Farley, (2001) *Leakage Management and Control: A Best Practice Training Manual*. World Health Organization. Geneva, Switzerland. London, 163 pp.

WATERPIPE Consortium. *WATERPIPE Newsletter*, no. 1. January, 2008.

Group, N.C. (2006) *International Water Survey & Cost Comparison*. NUS Consulting Group.

Chapter 3

Leakage monitoring and control: Pressure management, detection and location of leaks

Bambos Charalambous, Stuart Hamilton and Sophie Kanellopoulou

CONTROL OF LEAKAGE USING DISTRICT METERED AREAS (DMA)

The traditional approach to leakage control has been a passive one, whereby the leak is repaired only when it becomes visible. The development of acoustic instruments has significantly improved the situation, allowing invisible leaks to be located as well. But the application of such instruments over the whole of a large water network is an expensive and time-consuming activity. The solution is a permanent leakage control system whereby the network is divided into District Metered Areas (DMAs) supplied by a limited number of key mains, on which flow meters are installed. In this way it is possible to regularly quantify the leakage level in each DMA so that the leakage location activity is always directed to the worst parts of the network.

An important factor in lowering and subsequently maintaining a low level of leakage in a water network is pressure control. The division of the network into DMAs facilitates the creation of a permanent pressure control system, thus enabling pressure reduction in DMAs which reduces the level of background leakage, the rate of flow of individual bursts and the rate of the annual burst frequency.

Many water distribution networks are managed without using DMAs. However those that have successfully achieved low leakage levels without DMAs tend to have a combination of high quality infrastructure in good condition, an efficient repair operation and low, stable pressures.

THEORY OF DMA MANAGEMENT

The key principle behind DMA management is the use of flow to determine the level of leakage within a defined area of the water network. The establishment of DMAs will enable the current levels of leakage to be determined and to consequently prioritise the leakage location activities. By monitoring flows in the DMAs it will be possible to identify the presence of new bursts so that leakage can be maintained at the optimum level. Leakage is dynamic and whilst initially significant reductions could be made, levels over a period of time will tend to rise unless on-going leakage control is carried out. DMA management should therefore be considered as a method to reduce and subsequently maintain a low leakage level in a water distribution network. The key to DMA management is the correct analysis of the flow to determine whether there is excess leakage and identify the presence of new leaks.

Real losses are the difference between the system input and the total customer consumption (corrected for measurement inaccuracies) in a defined area. This is made up of Leakage (from mains, services up to the point of consumption and

 Integrated High Resolution imaging Radar and Decision Support System for the rehabilitation of WATER PIPElines. Edited by Matthaios Bimpas, Angelos Amditis and Nikolaos Uzunoglu. ISBN: 9781843393719. Published by IWA Publishing, London, UK.

storage tanks) and Overflows (mainly from storage tanks). Traditionally real losses were quantified as a volume and were calculated on an annual basis. However, this approach does not allow the necessary fine control of leakage to be achieved as it can take several months for a major change to be identified and the precision of leakage measurement is poor.

The extent of leakage can be gauged by assessing the 24-hour flow pattern of a network. A limited variation between the minimum and peak flow, particularly in a network with little industrial night use, is indicative of a leaky network. However this approach does not allow the leakage level to be directly quantified. Leakage is most accurately determined when the customer consumption is a minimum, which normally occurs at night. The size of the DMA will influence the level of burst leakage that can be identified. A large DMA will tend to have more leakage and customer night use, which will mean that a burst represents a smaller percentage of the minimum night flow, thus reducing its definition.

DMA LITERATURE REVIEW

Farley and Trow, 2003, referring to the DMA concept stated that "the technique of leakage monitoring is considered to be the major contributor to cost-effective and efficient leakage management. It is a methodology that can be applied to any network."

Dividing the distribution network into a number of leakage monitoring Zones or DMAs has been proved to be an efficient and effective technique for monitoring and controlling water loss. In Manual no. 4 of Managing and Reducing Losses from Water Distribution Systems, 2002, is stated that "dividing a distribution system into sectors such as PMZs and DMAs, which are then individually managed and monitored, enables:

- Continuous monitoring and recording of the total quantity of water entering each sector.
- Leak detection teams to target sectors and minimise the length of time leaks run for, if total supply is inexplicably high in particular sectors.
- Water quality incidents, for example dirty or discoloured water, to be contained within one sector and not spread throughout the network."

The application of the DMA concept has universal application. Irrespective of geographical location or distribution network size it has been established that this technique is reliable for leakage reduction and monitoring. However, it is important in order to maintain results to achieve long-term sustainability of the DMA concept.

MacDonald and Yates, 2005, stated that "the experience of DMA implementation at Halifax Regional Water Commission has been nothing but success".

Sturm and Thornton, 2005, reported that on the basis of their research project findings on nine North American Utilities with the goal to assess how international best practices can be transferred to North America, DMA techniques are applicable to Utilities in North America.

On the issue of sustainability, especially in developing countries, Loveday and Dixon, 2005, stated that "DMA implementation is a key component of NRW reduction activity and their maintenance must be seen as a long-term commitment".

DISTRICT METERED AREAS

Definition

A District Metered Area (DMA) is defined as an area of the supply network having ideally about 2000 properties supplied preferably from a single entry point which is metered (water entering and leaving) and pressure controlled.

Objective

The main objective of establishing a DMA is to reduce real losses to an economically acceptable level and to maintain this level through the application of proactive strategies, such as Active Leakage Control.

Advantages

There are a number of advantages in setting up DMAs, namely:

- The areas of the network are smaller, more manageable.
- The application of active leakage control is easier.
- Leaks are identified quicker based on MNF monitoring.
- Run time of leaks is much shorter.
  - Better pressure optimisation.
  - Lower water losses.
  - Financial savings.

Disadvantages

Minor problems may be encountered in the formation of DMAs which of course with proper planning and design can be resolved.

These problems are:

- Possible water quality problems associated with "dead ends" in the network
- Potential customer complaints due to optimisation of water pressures
- Difficult to be realised and maintained in complex urban networks

Design criteria

Key factors for good DMA design (Water Loss Task Force, 2007) are:

- Minimum variation in ground level across the DMA
- Easily identified boundaries that are robust
- Area meters correctly sized and located
- Single entry point into the DMA
- Pressure optimised to maintain standard of service to customers

PRESSURE MANAGEMENT

Management of pressure is a key factor in an effective leakage management policy. This has long been recognised by many utilities and the ultimate goal is for all District Metered Areas (DMAs) to be equipped with Pressure Regulating Valves (PRVs) in order to reduce pressure in the network to the minimum acceptable levels and where reduction is not practicable to control and stabilise pressure.

Measurements of pressures within the DMAs are usually carried out to establish operating pressures at the low, medium and high points of a DMA as well as the Average Zone Night Pressure (AZNP) for the DMA. Furthermore, the pressure measurements are critically examined with the aim to reduce pressure as much as possible whilst maintaining the minimum level of service to the consumers at the critical point at maximum demand.

Pressure management does not only involve pressure reduction but also other methods of controlling and optimising pressure in order to achieve the maximum possible pressure reduction without compromising customer service and minimising leakage at the same time. A definition of pressure management in its widest sense has be given by Thornton *et al.*, 2005 as "the practice of managing system pressures to the optimum levels of service ensuring sufficient and efficient supply to legitimate uses and consumers, while reducing unnecessary excess pressures, eliminating transients and faulty level controls all of which cause the distribution system to leak unnecessarily."

According to Farley and Trow, 2003, "there are several benefits in applying pressure management, and if it is designed and maintained well, there are few, if any disadvantages". Water utilities will therefore only benefit from the adoption of an effective pressure management policy which inevitably will have positive results in improving the operational performance of the network and in providing better customer service. International experience has shown that there are several benefits (Farley and Trow, 2003) in adopting an appropriate pressure management police, such as:

- Reduction of leakage – bursts and background
- Reduction of pressure – related consumption
- Reduction of frequency of bursts
- Provision of more constant supply to customers

The above benefits are being recognized worldwide and the number of utilities applying pressure management is continuously increasing. Thornton and Lambert, 2005, international experts in pressure management, stated that "an ever-increasing number of countries and utilities are now recognizing that good pressure management is the fundamental foundation of good leakage and infrastructure management".

It is common practice for water utilities to design their distribution networks to provide the minimum standard of service to the customers who are at the highest location of the system at maximum demand. This means that the minimum pressure occurs at this location at some time during the day when demand is at its maximum. Depending on how well designed the network is, significant variations in demand are reflected in varying pressures ranging from minimum at maximum demand to maximum at minimum demand at the critical point.

Understanding this concept is of the utmost importance since regulation of pressure can bring about significant reduction in leakage without compromising the level of service to the customers. McKenzie and Wegelin, 2005, based on the results of a project which they carried out in Sebokeng/Evaton, South Africa, reported that "by controlling pressures during the off-peak periods it was often possible to significantly reduce the losses without identifying or repairing a single leak. After the excessive pressures have been addressed, the other measures such as repairing pipes and /or retrofitting can be tackled".

The Water Loss Task Force promotes the adoption of the "four component" approach for managing real losses. Figure 3.1 shows that Pressure Management has an influence on the other components with favourable results, namely:

- Water conservation
- Reduced pipes failures
- Extended asset life
- Cost savings

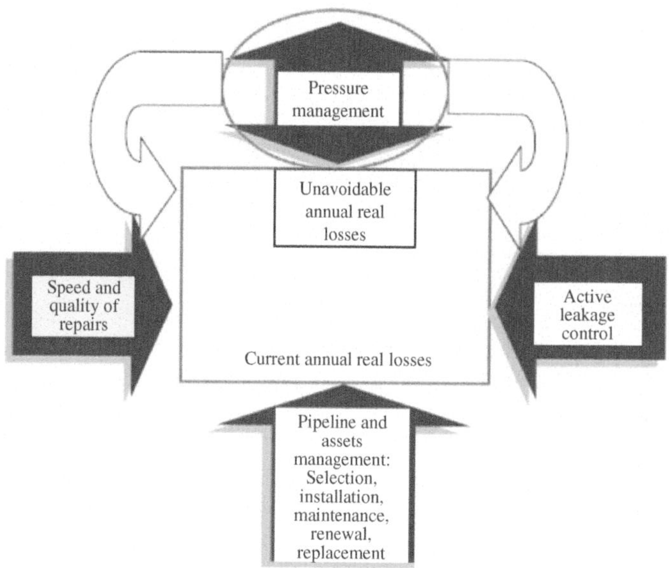

Figure 3.1 The four component diagram with secondary influences of pressure.

Understanding the impact that pressure management will have on the operation of a distribution system is essential in the application of the correct type of approach and equipment. Pressure management is the optimisation of pressure in a distribution system to provide the minimum level of service to all customers, which ideally should be around 20m with absolute minimum of 15 m if conditions allow.

Pressure management is achieved through pressure control devices, such as pressure reducing or regulating valves. The types of pressure management that are most commonly used are:

- Fixed outlet
- Two point control (Time or Flow)
- Flow modulation

In the case of a fixed outlet pressure regulating valve (PRV), the pressure is continuously regulated immediately downstream of the PRV irrespective of the value of the pressure in the network. The pressure at the PRV is usually set so that the minimum level of service is achieved at the critical point at maximum demand. The draw back of this particular method is that the pressure in the network rises during periods of minimum demand without effectively being able to apply further control.

The other two methods provide better control of pressure. Two point control PRV is used to regulate the pressure in the network at two different pressures depending on the demand for flow or the time period. Again the setting of the pressure is on the downstream side of the PRV at a fixed value. Flow modulation is the most advanced method of pressure regulation and it's increasingly gaining ground as the benefits from using such a method are becoming known and are backed up by tangible evidence form field applications worldwide. With this method, pressure is continuously controlled based on the demand so that at the critical point in the network the pressure is always maintained at the minimum level of service thus, achieving maximum pressure reduction at any time.

EFFECT OF PRESSURE

The effect of pressure, though well understood in theory, has only recently been appreciated in the management of leakage, both in terms of reducing and maintaining a low level of leakage in a water network. In essence, the simplest and most reliable expression for the relationship between pressure and leak flow rates is:

$$L_1 = L_o(P_1/P_o)N_1$$

where P_o and L_o are the initial pressure and leak flow rate in the network, P_1 and L_1 are the values at a revised pressure, and the exponent N_1 typically varies from 0.5 to 1.5, for individual DMAs, depending upon the predominant type of leaks, and whether pipe materials are rigid or flexible.

The average N_1 value for large systems with mixed pipe materials is often assumed (for simplicity) to be 1, implying a linear relationship between leak flow rates and pressure.

The pressure in a network varies according to the flow. So as the flow increases e.g. at times of peak demand) the pressure will reduce, thus reducing leakage as shown in Figure 3.2.

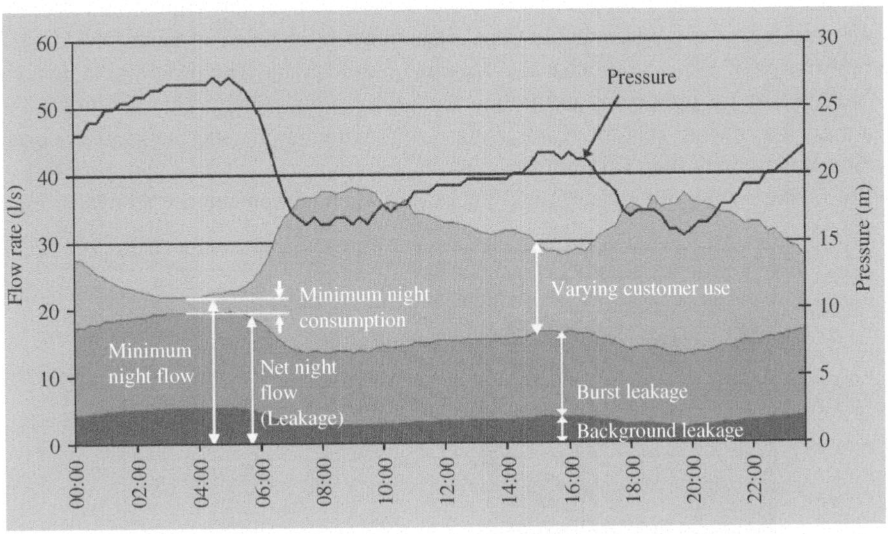

Figure 3.2 Variation of flow, pressure and leakage in a water network.

Accordingly, the leakage rate is not constant over a 24-hour period. The parameter used to relate the night leakage rate to the daily leakage rate is known as the 'Night-Day Factor' (NDF) and is determined as follows:

- Daily leakage volume = (NDF) × (night leakage rate per hour)

where NDF is expressed in hours per day.

Therefore the daily savings equals to $(L_1 - L_o\ (P_1/P_o)) \times (NDF)$

For DMAs with gravity supply, the NDF is usually less than or equal to 24 hours per day, and for low pressure gravity systems with large frictional head losses, the NDF can be as low as 12 hours per day. However, for DMAs supplied by direct pressure, or those with pressure modulation devices (based on time, or flow), the NDF is usually greater than 24 hours per day, and can be as high as 36 hours per day.

The NDF is clearly a major factor which must be considered when using night flows to estimate daily and annual leakage. For this reason, it is always preferable to express leakage rates based on measured night flows on a 'per hour' basis, and leakage rates based on water balances on a 'per day' basis.

COMMUNICATION TECHNOLOGY

It is evident that times are changing and globalisation, information technology, innovation and sustainable development are part of our daily lives. Water organisations are considered to be extremely conservative especially when it comes to change. It has been the approach for many years that if something works, do not change it. However, a number of water companies recognised the need to go forward toward a higher standard of performance.

Many water utilities have used Information Technology (IT) to improve reliability and reduce costs. The need for change is becoming more and more pressing in order to facilitate and improve daily operations and to collect data for timely analysis and decision making as well as forward planning.

The experiences gained in the past from the application of SCADA systems prove invaluable when there is a need for implementing continuous monitoring of the flow into the District Metered Areas (DMAs) as well as of the pressure. Both flow and pressure are essential for the implementation of any leakage reduction strategy.

The selection criteria for the continuous monitoring system are decided taking into consideration present and future needs and requirements. To this end careful consideration and examination of the available techniques, methodologies and technologies is given in order to adopt an appropriate system with low capital expenditure, operation and maintenance cost.

The criteria that could be used are summarised below:

- Small capital expenditure with minimum maintenance
- Easy to install, simple and easy to operate
- Low running costs and easy to expand
- Continuous recording and storing of data on site
- Downloading of data on request or at preset times
- Multi communication capability via Public Switched Telecommunications Network (PSTN), Global System for Mobile Communications (cellular phone technology-GSM), Radio, World Wide Web (www)

Telematic type of solutions which combine information technology and telecommunication networks using the World Wide Web for data transfer are more and more preferred due to their low capital and operational costs.

The heart of such a system is a programmable controller which is installed at each DMA meter location and is capable of precise registration of events and optimal archiving of stored data. The programmable controller is installed inside a stainless steel cabinet for protection and is powered using solar energy. This solution provides an excellent performance to cost ratio. The overall cost for each station is approximately €1,800 but most importantly the operating cost of such a solution is extremely low, €10/month.

A typical layout of a DMA monitoring station and the cost of each component are shown in Figure 3.3.

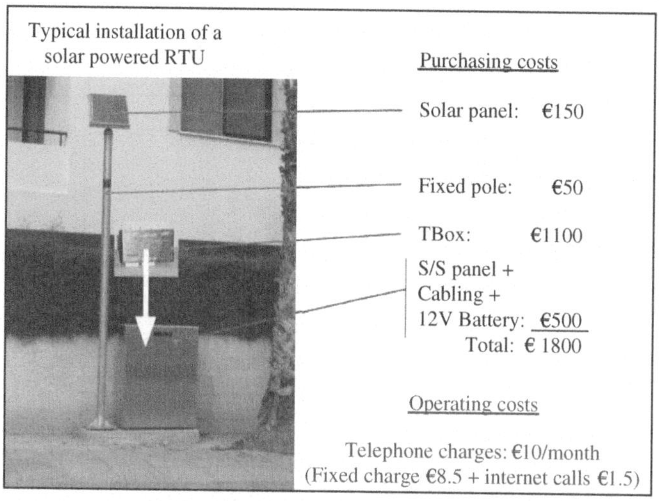

Figure 3.3 Typical layout of DMA monitoring station with cost breakdown.

The continuous monitoring of the DMA meters combines information technology and telecommunication networks to transfer the data via the World Wide Web. The historical data gathered in the programmable controller of each DMA is sent by the controller to an email account. Operating software installed in the dedicated computer at the utility's control room connects to this email account and downloads the data, which are first sorted according to the DMA and then are used to update existing reports. The monitoring system is shown diagrammatically in Figure 3.4. Direct access to the programmable controllers from the control room enables modification to the programming of the controllers, downloading of historical data on request and closing or opening of the PRVs.

The operating software is an intelligent communication interface between the monitoring station and the central control. It exploits the power of the internet by receiving data from the monitoring stations in the form of e-mail attachments. It is capable of transferring and exporting data to the majority of data bases and has powerful graphics to display data in the forms of curves and statistical tables (Figure 3.5).

There are several benefits to the Utility in applying the above DMA monitoring system. The operational cost of the system is very low and it basically requires no maintenance. The system provides an environmentally friendly solution utilising solar energy for power. It is considered to be the backbone of the leakage management strategy providing data on continuous basis which is used to analyse the water loss in the DMAs and set priorities for leak location. In addition it provides early warning for large bursts thus triggering the mechanism for leak location and repair.

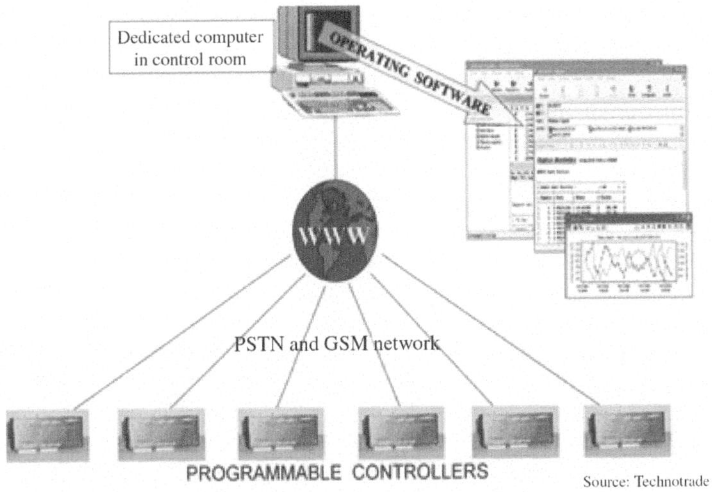

Figure 3.4 DMA monitoring system – electronic mail support.

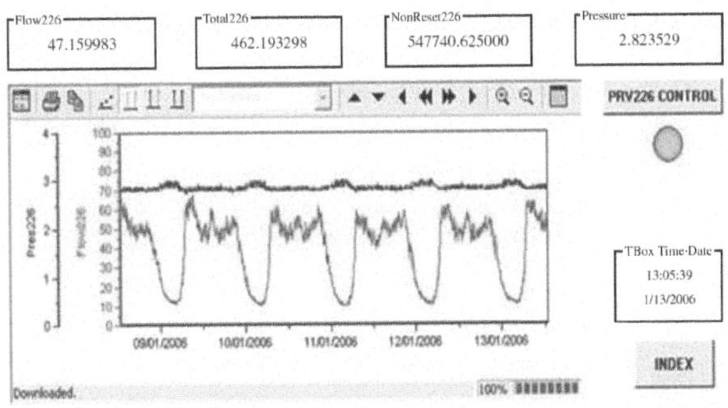

Figure 3.5 A typical flow and pressure monitoring template obtained from the system.

The continuous monitoring system is a valuable asset for the Utility providing all the necessary information for the efficient and effective management of the distribution network.

IMPLEMENTING PRESSURE MANAGEMENT

Continuous flow monitoring enables the establishment of flow patterns for the DMAs providing essential information such as maximum and average daily flows as well as minimum night flows. Data required establishing legitimate customer night use and background leakage in DMAs is also needed. Having available this information, the Burst and Background Estimates (BABE) component approach to leakage is used to analyse the Minimum Night Flow (MNF).

Measurement of pressures within the DMAs are carried out to establish operating pressures at the low, medium and high points of the DMA as well as the Average Zone Night Pressure (AZNP) in the DMA. Furthermore, the pressure measurements are critically examined with the aim to reduce pressure as much as possible whilst maintaining the minimum standard of service to the consumers. As a rule a minimum standard of service of 20 metres at the critical point in the DMA at maximum demand is usually considered. According to common practice, pressure should not be reduced in a single step otherwise there could be adverse effect on customer supplies. This of course is valid where the customers are directly fed from the mains. In systems where customer roof tanks are used, the effect on customer supplies should be minimal since all water used in the household comes from the roof tank. Figure 3.6 shows a fixed reduction in pressure which has as a direct result, a reduction in the Minimum Night Flow.

It is possible to optimise pressure further using advanced techniques, such as flow modulation, in order to achieve further reduction thus, driving leakage to even lower levels.

FLOW MODULATION

Flow modulation provides an advanced method of controlling pressures and the outlet pressure is continuously controlled and varied so that the pressure required at the critical point in the network is always maintained at acceptable levels. In this manner during periods of high demand, the valve adjusts itself to increase the flow in order to maintain acceptable pressures in the system. When demand in the system is reduced, the valve readjusts so that excess pressures are reduced thus reducing leakage further. The benefit can clearly be seen in Figure 3.7 below. It is evident that after installing a fixed outlet there was a reduction in the Minimum Night Flow (MNF) of 4.1 m^3/h. When flow modulation was effected, there was a further reduction of 2.5 m^3/h, which proves that provided the conditions are such that favour the application of flow modulation, further reduction in leakage is possible.

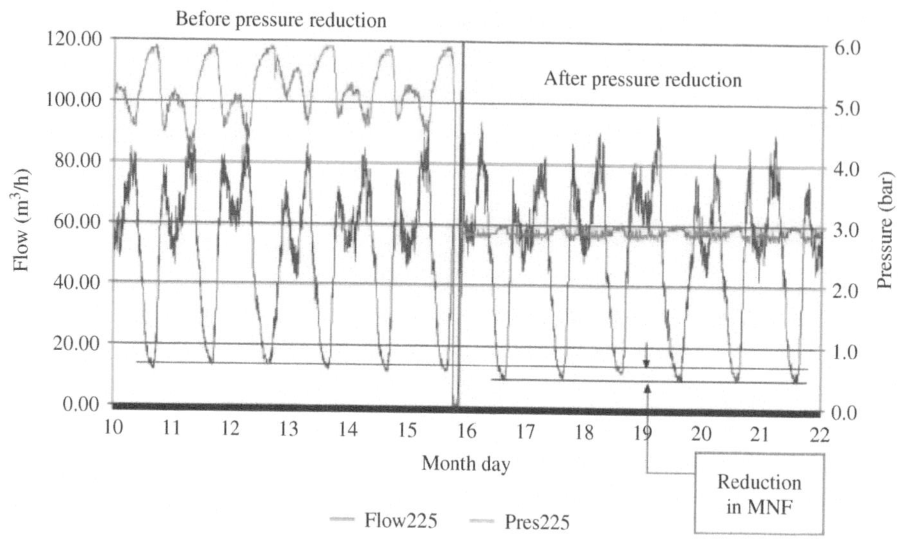

Figure 3.6 Pressure reduction.

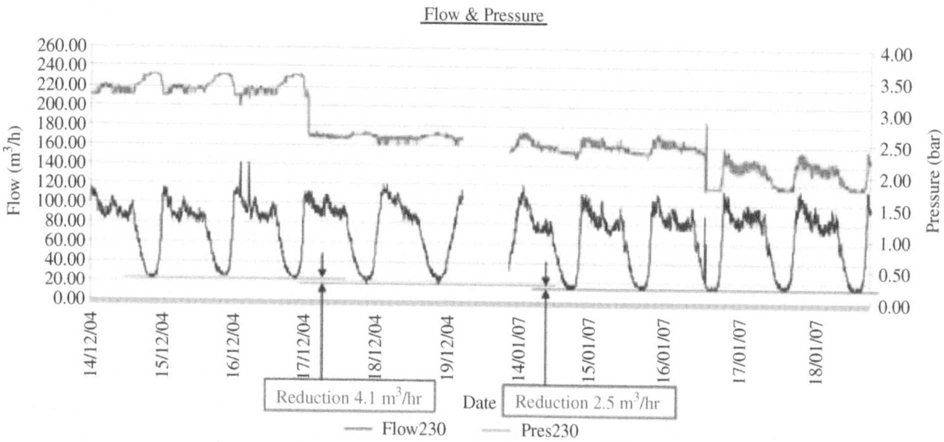

Figure 3.7 Application of fixed outlet and flow modulation PRV.

LEAKAGE CONTROL USING MINIMUM NIGHT FLOW (MNF)

The Minimum Night Flow obtained from the continuous monitoring of flow is plotted over time and is used as a tool in monitoring and controlling leakage. Figure 3.8 shows a normal plot of the MNF with only seasonal variation showing an increase in the summer months. On the other hand Figure 3.9 shows an increase in the MNF with time which must be investigated using leak detection and location equipment.

Of course in the event of a large leak in the system this immediately becomes apparent on the MNF plot as shown in Figure 3.10. This way the Leak Run Time is kept to a minimum thus, reducing the volume of water lost. Having a MNF monitoring, it takes only two to three days to become aware of the leak and with a well equipped and trained staff it is a matter of days to locate the most difficult of leaks. Equally important is the time taken for repair of the leak which again must be kept to the minimum possible.

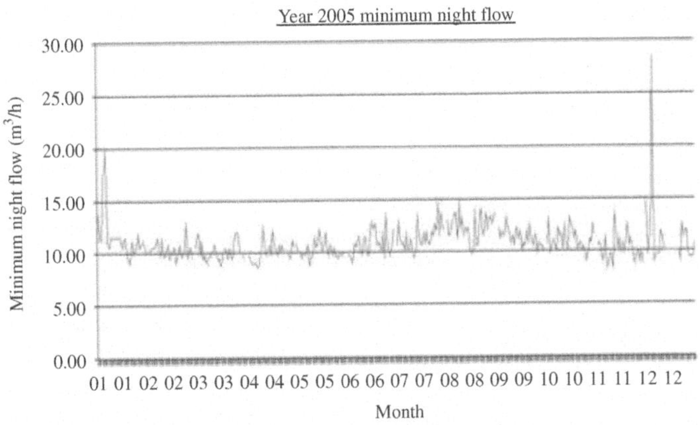

Figure 3.8 Typical MNF.

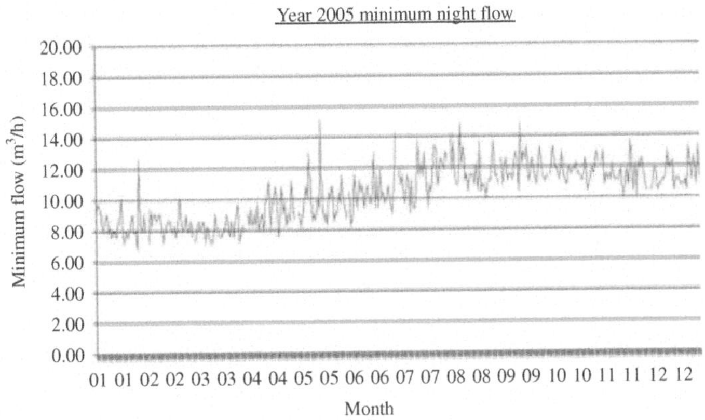

Figure 3.9 Increasing MNF.

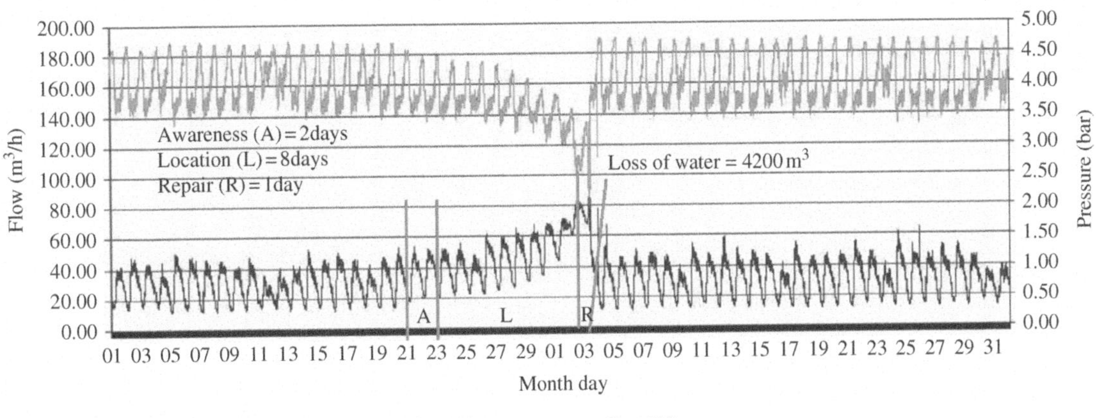

Figure 3.10 Leak Run Time – April–May 2004 Flow & Pressure.

REFERENCES

Charalambous, C.N. (2003) "Use of Telecommunications in Water Distribution Systems", The 9th EATA International Conference on Networking Entities – Technologies for the Future, Lemesos, Cyprus, Proceedings, pp. 121–134.

Charalambous, C.N. (2005) "Experiences in DMA redesign at the Water Board of Lemesos, Cyprus", Specialised Conference on Leakage, Halifax, Nova Scotia, Canada. Proceedings pp. 403–413.

Farley, M. and Trow, S. (2003) Losses in Water Distribution Networks, A Practitioner's Guide to Assessment, Monitoring and Control, London, IWA, ISBN1 900222 11 6, pp. 146–149.

Fanner, P. (2004) Assessing Real Water Losses: a practical approach. Water 21 April 2004, ISSN1561–9508, IWA.

Lambert, A.O. (2001) What do we know about pressure: Leakage Relationships in Distribution Systems? IWA Conference on system Approach to Leakage Control and Water Distribution systems Management, Brno ISBN 80-7204-197-S.

Lambert, A.O. and Fantozzi, M. (2005) Recent Advances in alculating Economic Intervention Frequency for Active Leakage Control and implications for calculating of Economic Leakage Levels. IWA International Conference on Water Economics, Statistics and Finance, Crete, Conference Proceedings.

LEAKS suite of software, ILMSS Ltd.

Loveday, M. and Dixon, J. (2005) DMA Design Sustainability in Developing Countries. IWA Specialised Conference "Leakage 2005", Halifax, Nova Scotia, Canada, Conference Proceedings, p. 252.

MacDonald, G. and Yates, C.D. (2005) DMA Design and Implementation, a North American Context. IWA Specialised Conference "Leakage 2005", Halifax, Nova Scotia, Canada, Conference Proceedings, p. 175.

Mckenzie, R.S. and Wegelin, W. (2005) Sebokeng/Evaton Pressure/Leakage Reduction: Public Private Partnership, IWA Specialised Conference "Leakage 2005", Halifax, Nova Scotia, Canada, Conference Proceedings, p. 385.

Manual 4, Toolkit, (2002) Establishing Pressure Management Zones and District Metered Areas: State of Queensland, Environmental Protection Agency, Wide Bay Water, ISBN 0 7242 9493 7, p. 7.

Sturm, M. and Thornton, J. (2005) DMA Proactive Leakage Management using DMA and Pressure Management – Is it applicable in North America. IWA Specialised Conference "Leakage 2005", Halifax, Nova Scotia, Canada, Conference Proceedings, p. 426.

Task Force on Water Losses (2000) Losses from water Supply Systems: Standard Terminology and Recommended Performance Measures, IWA.

Thornton, J. and Lambert , A. (2005) Progress in Practical Prediction of Leakage: leakage, pressure: burst frequency and pressure: consumption relationships, IWA Specialised Conference "Leakage 2005", Halifax, Nova Scotia, Canada, Conference Proceedings, p. 347.

Thornton, J. Shaw, M. Aguiar, M. and Liemberger, R. (2005) How Low Can You Go? A Practical Approach to Pressure Control in Low Pressure Systems, IWA Specialised Conference "Leakage 2005", Halifax, Nova Scotia, Canada, Conference Proceedings, p. 392.

Water Loss Task Force, DMA Guidance Notes, IWA WLTF, 2007.

Chapter 4

Equipment for leak detection and location

Gordon Watt, Alan Jones and Stuart Hamilton

WHAT TECHNOLOGIES MAY FIND LEAKS IN 'DIFFICULT' PIPES?

In the toolbox of available leak detection technology, there are several "conventional" technologies for localizing, locating and pinpointing leaks, in both the distribution network and transmission mains. All these technologies work well on 'conventional' pipes – small to medium diameter, metallic pipes at reasonable pressure. They are:

- Leak noise correlator
- Correlating noise loggers
- Gas injection and detection at the leak point

But every network has a proportion of large diameter, non-metallic pipes, such as Polyvinyl Chloride (PVC), medium density polyethylene, asbestos cement or reinforced concrete. These are usually transmission mains, frequently in low-pressure systems. Such mains can traverse many kilometres of rural terrain, will have few valves or other suitable contact points for sensors and hydrophones, and will often be laid in positions where they are strategically difficult to monitor and where excavation cost is at a premium – for example beneath a river bed or a main highway. They are usually laid deeper than distribution pipes and depths of 5.0 metres or more are not uncommon. In such cases accuracy of leak location and pinpointing is paramount to avoid major disruption and unnecessary excavation costs.

Leaks in these pipes are difficult to locate with conventional technology, and even the most advanced leak noise correlator will require the main to be excavated at say, 500 m intervals, for the installation of sensor contact points or hydrophone insertion points, to maximize the leak noise. In-pipe acoustic technology is able to pinpoint the leak very accurately, but the cost of excavation of the insertion point and chamber, particularly for deep laid pipes, is high.

There are, however, some less conventional techniques, which are invariably called on when the other technologies fail:

- Correlation with low frequency hydrophones
- Signal analysis at low frequencies
- In-pipe acoustic technology
- Ground penetrating radar (GPR)

Of all the technologies, only one is a truly non-intrusive technology – GPR. If this technology can be enhanced to distinguish bursts and leakage from other underground anomalies, we will have solved the quest for the perfect leak detection technology. The advantages and disadvantages of both the "conventional" and the "new" technologies are discussed in the following sections.

THE "CONVENTIONAL" TECHNOLOGIES

The leak noise correlator – established technology with new features

The utility operators of advanced utilities have a wealth of acoustic tools and equipment at his/her disposal with which to find bursts and leaks. The leak noise correlator, instead of depending on the noise level of the leak for its location, relies on the velocity of sound made by the leak as it travels along the pipe wall towards each of two microphones placed on conveniently spaced fittings. Hydrophones, placed in the water column, can also be used to enhance the leak sound in plastic pipes or large pipes.

Once considered the ultimate leak location tool, the correlator has now been around for in excess of 20 years, gradually changing in size from a unit that required a vehicle to carry it, to the modern range of portable hand-held units. There is no doubt that the latest versions of the correlator, with the capability of frequency selection and filtering, can quickly and accurately locate a leak (to within 0.5 metres) in metallic pipes of small diameter. The modern instrument has several advantages:

- It is portable and only requires a single operator;
- There is now a range of models of varying levels of sophistication;
- Some models are available at very low prices, and still do an adequate job for most situations;
- Personal Digital Assistant (PDA) based leak noise correlators are compact enough to be carried around the world in a lap top case;
- A 'Bluetooth' portable communication link can communicate with the sensors/transmitters by digital radio, so the distances between the sensors/transmitters is no less than conventional correlators;
- If used as a noise logger and left overnight the sensors can be configured to perform 4 correlations within a 24 hour period;
- Units generally use three sensors as standard for increased accuracy in triangulating the exact velocity without having to generate a temporary 'leak'. This facility has proved popular with leak detection contractors, who welcome fully digital correlators with a three-sensor system as a means to finding more difficult leaks.

However, the correlator requires contact points at frequent (500 meters or so) intervals for attachment of the sensor. This is a disadvantage for leak detection in transmission mains, where the main has to be excavated at intervals and contact points installed. Hydrophones can be used to enhance low frequency noise, but these also require insertion points to be made in selected under-pressure tapping points.

But one UK leak detection contractor quotes the success of finding a large leak in a 900 mm ductile iron main after responding to a call from a major construction company based on the Isle of Man. The main was 2.5 metres deep, 1,500 metres in length, and laid at the side of a major highway in Douglas, the island's capital. Because of the nature and position of the main, it was imperative that the correlation was accurate and precise. The main had failed a pressure test and water loss was estimated between 1.0–3.5 litres/minute. Water was not showing on the surface. The distance between fittings for correlation purposes was 470 metres and 985 metres respectively. The first correlation of 470 metres produced a peak – this was measured out on site and the position marked. Several other correlations were conducted and the position was confirmed. Hydrophones were utilised to confirm the position. The process was repeated for the second correlation of 985 metres and the position was marked on site. On the first correlation the leak was located within 0.5 metres of the marked position and within 1.0 metres on the second correlation.

Leak localizing – a new survey tool

In recent years, the correlator has been complemented by the technique of leak localising – or noise logging – a process for narrowing down areas of a distribution network or a DMA which contains a burst or number of leaks. The technique – also widely used for detecting leakage in transmission mains – has become increasingly popular and the technology has moved on a pace. A cluster of loggers, usually 6, 8, 15 or more depending on the manufacturer, is deployed in an area to be surveyed, each logger placed on a hydrant, meter or other surface fitting. Logger data can be analysed insitu, either by downloading on site, or by driving past each logger with a van-mounted receiver to pick up anomalous noise signals. Noises which are suspected of being leaks can be confirmed, and the leak is located using location equipment such as the leak noise correlator, an acoustic 'mat', ground microphone or acoustic sounding stick.

Correlating noise loggers

The latest generation of noise logger systems now incorporates a multi-point correlation facility to provide 'instant' location of the leak position – the correlating logger – a facility which is particularly relevant to transmission mains.

New versions of correlating loggers, with enhanced software to increase the speed of data analysis, are continually being developed. Correlating noise loggers can be enhanced by the addition of a low cost and low power Global System for Mobile Communications (GSM) engine. As with other types, the loggers are normally be installed in clusters and used in two basic modes. The first mode is as a noise recorder where the data is sent from each logger to a host personal computer PC (normally each day) – or to the Internet – for analysis. The second mode allows the logger to determine if a leak is present and a Short Message Service-SMS (text message) is sent along with the data if a leak occurs. GPRS transmission can be used for data transmission and SMS for alarms. The noise data can be used to correlate between loggers to identify the leak location. Correlation between adjacent loggers along a length of transmission main, or as a survey tool, is another useful addition to the multi sensor toolbox.

Gas injection

Although not a new technology, the potential of leak detection using gas technology is not always fully realized. This is another contractor-service technology, and it is another addition to the toolbox for finding difficult leaks. It is particularly successful for finding small leaks in service pipes where the volume of gas required is small.

In Gothenburg, the water utility has used the technique for the last twenty years, for finding difficult leaks in pipes of all materials, but particularly for finding small leaks on new plastic pipes which fail a pressure test. The gas used is "Formier 10", ten per cent hydrogen and ninety per cent nitrogen, of which hydrogen is the active part. Hydrogen, the lightest of all gases, has a unique capability of escaping through very small fissures. In buildings, Formier 5 (five per cent-5%- hydrogen) is used. The time taken for the gas to reach the surface depends on the soil conditions, depth of the pipe and the size of the leak. The gas detector is sensitive to very small quantities of gas, and is able to detect very small leaks. However, for large low pressure mains, the high volume of gas required rules out using this technology in live mains, except in exceptional circumstances.

THE 'NEW' TECHNOLOGIES

Correlation using low-frequency hydrophones

Using the correlator on large diameter transmission mains has long been recognised as a difficult application. All correlator manufacturers are able to quote some success, but agree that better filtering and sensor technology is required to overcome the poor sound propagation qualities of trunk mains and large distances between access points. Very special sensors are required to operate on a far more focused frequency spectrum. However, manufacturers are now improving the efficiency of a leak noise correlation by enhancing the technology with low frequency hydrophones.

There have been some success stories with low-frequency correlation. One manufacturer has developed a low frequency hydrophone specifically for large diameter mains and has also changed the software of the analysis program for this application. The Figure 4.1, illustrates such a story.

Chicago case study

In September 2007 the City of Chicago water utility was called informing them that water was entering the basement of a school in Dearborn street in downtown Chicago. A contractor was called in to find the leak, using a correlator and low frequency hydrophones.

In Dearborn Street, there are two 8 inch cast iron water mains and one 36 inch reinforced concrete water main. Correlations of the 8 inch mains did not indicate any leakage and therefore investigation of the 36 inch main was scheduled. Hydrophones were installed – directly on to a 1 inch pitot tube tapping point at one end of the pipe section, and on a fully charged hydrant on a 12 inch connection off the main pipe at the other end. The leak was not obvious at first due to an out-of-bracket noise (probably pump noise), but by filtering this noise out of the correlation, the engineers were able to determine the presence of a leak on the main. Actual measurement of the pipe between the two hydrophones including the hydrant leg, took place using a measuring wheel, indicating that the actual length to be 829ft. It was also necessary to measure the velocity of sound in the concrete pipe, and this was done by creating a 'leak' outside of the bracketed area by opening up a hydrant and recording the time delay and velocity of the out-of-bracket sound. Entering the pipe length and pipe velocity into the calculation, indicated a leak pinpoint of 408ft from one of the hydrophones station. The graph in Figure 4.1 shows the screen produced by the software, indicating the correlation peak.

The measurement point lay directly over a known concrete to ductile iron adaptor in a repaired section of the concrete pipe. Leak pinpointing was done with a ground microphone. The leak was the result of a torn gasket on the adaptor, probably done during original construction.

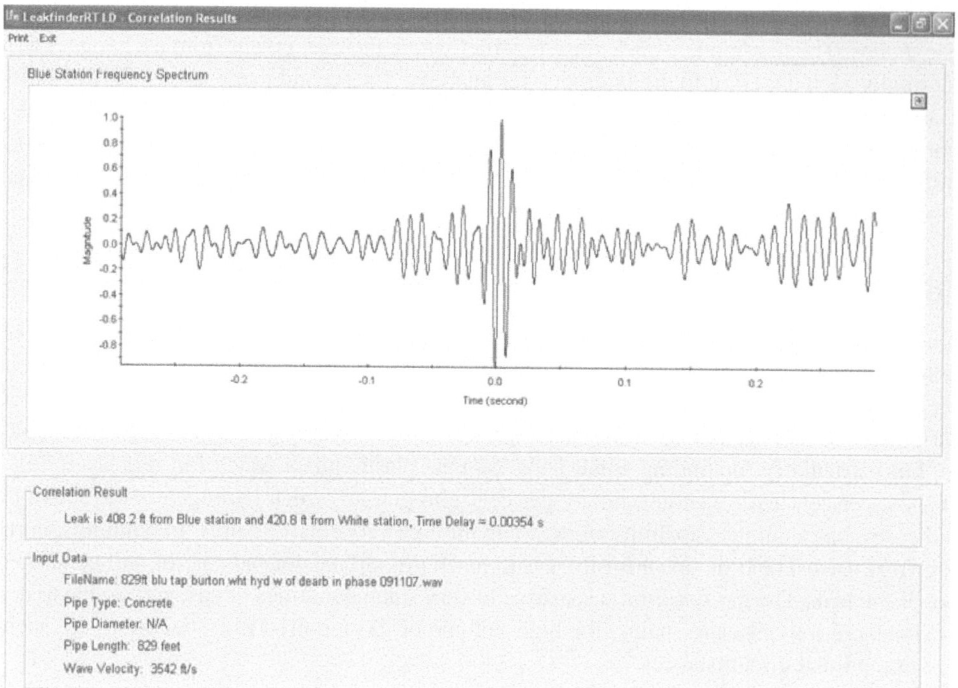

Figure 4.1 Correlation using low-frequency hydrophones (Source: Echologics/ADS).

Signal analysis

Designed specifically for detecting 'difficult leaks' in plastic pipes, this concept is relatively new to the market. It was first publicized at Leakage 2005 (3) when the concept of signal analysis to frequency shift ('pitch up') signals transmitted by a low frequency leak was discussed. The frequency shift principle is a well established practice for tracing buried plastic pipes: a pulse emitting device is attached at the pipe's accessible point, after which the pipe can be traced by means of a ground microphone that picks up the pulses propagating through the water in the pipe. The frequency-dependent attenuation is similar to that of plastic pipes above the surface, and an "Ultrac Leak Finder" (ULF) coupled to the ground microphone output will extend the range of detection from a normal 10–20 metres up to at least 300 metres. At the same time the pulse power can be held very low, thereby avoiding risks of damage to the pipe.

Based on this principle, the 'Ultrac Leak Finder' instrument was shown at the Global Leakage Summit (4) in January 2007 and is now commercially available. At present the instrument can only locate leaks to a section of pipe, and is unable to pinpoint the exact leak position. The manufacturer has been working on this gap in the technology, and is now refining the technology for locating the leak more precisely, confirming it with a large number of field trials. A pinpointing accuracy of ±5.0 metres is currently being achieved.

In-pipe acoustic technology

A technology which is fast becoming a support tool to the correlator for finding leaks in this type of main is in-pipe acoustic location, with the sensor inserted into the pipeline via a cable or via a free-swimming plastic ball.

The Sahara

The only commercially available model of cabled equipment is the 'Sahara', provided as a contracting service by WRc (UK) and the Pressure Pipe Inspection Company (Canada) to utilities across the world. Figure 4.2 shows the arrangement for using the equipment. A microphone is inserted into a pressurised main via an air valve or an under-pressure insertion point in a specially constructed chamber. The microphone cable is calibrated to measure the distance from the entry point to the leak, which is identified and recorded as the microphone passes by. The signal is picked up by a "Pipespy" receiver which has a 4.0 m range and 100 mm accuracy. Leak print-outs can be recorded photographically or via GPS. Because the sensor hears the leak noise from inside the pipe, the equipment is suitable for all types of material. Although expensive per inspection chamber (installation and inspection cost), the equipment comes into its own for finding leaks where other techniques are not possible, such as strategically-laid pipes under main highways and rail tracks and under house basements.

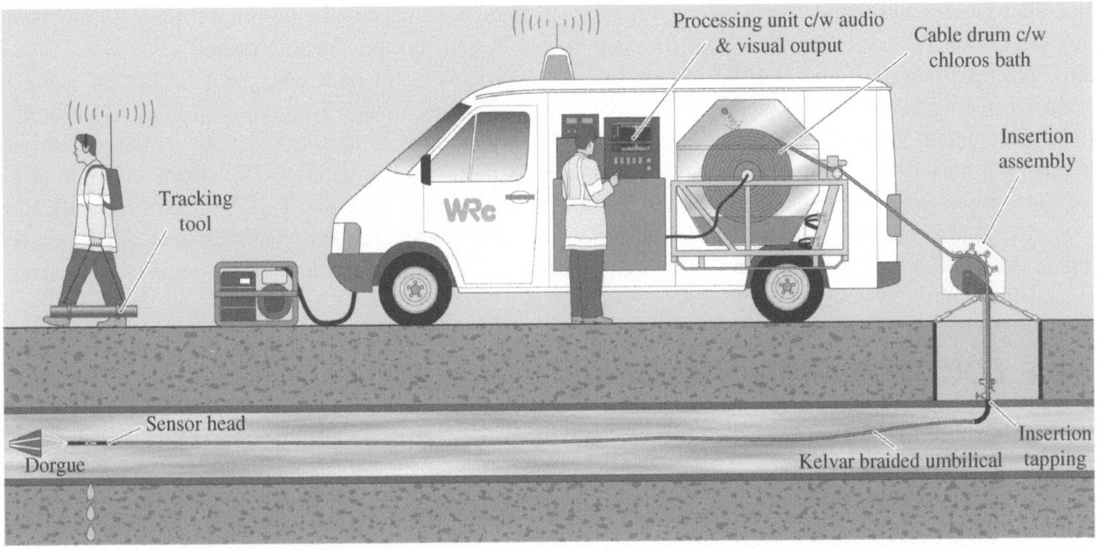

Figure 4.2 Schematic showing Sahara in-pipe acoustic technology (Source: WRc Plc).

Several Sahara case studies, with costed results, were reported at Water Loss 2007, Bucharest (5) . One of these is the Thames Water 'trunk main leakage programme', which began in 1998. 1500 surveys have been carried out and 1250 leaks found, the average distance between leaks being 505 metres. During the first two years of the survey the utility reported that water loss from the transmission main network had been reduced by 65 Mld. The leaks ranged in size from 0.5 Mld to 10 litres/hour.

Wireless technology

Another technology that has been developed is a neutrally buoyant ball, which, like the Sahara, also enters a pipeline under pressure via a valve. One instrument, the "Smart Ball" is a free-swimming foam ball with an instrumented aluminium core, capable of detecting very small acoustic events in pipelines. Another, still under development, is the 'Urchin' leak detector – a 50 mm diameter plastic ball. Using wireless communications, it listens for leaks as it travels along a pipeline. Inserted into a main through a standard fitting, it moves with the water flow, recording the positions of any leaks as it passes. The device has a wireless communication system with the surface that enables regular checks on position to be made along the route. When fully developed, location accuracy is expected to be within a few metres – an acoustic mat or other location device would be used to home in on the exact position.

WHAT IS GPR AND HOW DOES IT WORK?

Ground Probing Radar, sometimes called Ground Penetrating Radar, is a non-intrusive investigation tool used for locating buried utility infrastructure such as pipes, cables, manholes etc without digging.

The name RADAR is short for Radio Detection and Ranging. The object enables the direction and distance of a target to be calculated from the point at which a radio signal of a known frequency has been transmitted.

In the case of Ground Probing Radar; waves of Electromagnetic (E/M) energy are pulsed into the ground from an antenna placed on, or just above the ground surface. These pulses of energy interact with the various materials encountered causing some of that energy to be reflected back to the surface. Differences in the properties of those materials: dry, wet, hard, soft, conductive, non-conductive etc cause subtle changes in the energy that is reflected. As the frequency at which the waves of energy are being pulsed into the ground is known, it is possible to calculate the distance from the transmit and receive antenna position to the point where those subtle changes occurred. Because the receiver has a directional antenna, it is also possible to calculate the angle or bearing from which those changes emanated. By repeating this process many times from many locations only millimetres apart, the position of those changes and therefore the position of the material changes that caused them, can be mapped with high approximation using basic trigonometry.

HOW DOES GPIR DIFFER FROM GPR?

With a normal GPR system, the collected data is processed and viewed 2-dimensionally, most often in cross section, but also sometimes in plan view. Although representative of the data collected, to the uninitiated it is often difficult to visualise the true size and shape of any detected feature. Its main advantage is the minimal time requirements for data processing. It is often the preferred choice for the mapping of buried services for this reason.

Ground Probing Imaging Radar (GPIR) systems take the processing of data one stage further by using modern 3-dimensional imaging techniques to create a point cloud or wire frame model from which a more visually indicative likeness can be created. This is the preferred option for those interested in the assessment of asset condition as the displacement of a buried service and any change over time is more readily spotted. The down side is that a higher density of data points must be collected in order to build an accurate 3d model. This increases the collection time required on site. It also increases the time required for post processing of data off site. This time difference is slowly shortening as the availability of modern low cost/high specification computers with large amounts of processing power and high quality graphics systems is improved.

GPIR PIPE LOCATION METHODOLOGIES

When used for locating material changes that are linear in nature such as a buried water pipe, two typical methods are employed:

Where the direction of the pipe is already known; the GPIR system can be moved across its path at right angles to it. This optimises the reflected changes emanating from the pipe. By repeating this process at various locations along the known route, the approximate location and depth of the pipe can be determined.

Where the direction of the pipe is uncertain, or where several pipes may be in close proximity, it is more appropriate to establish an orthogonal grid. This is built from a number of equally spaced reference points along a series of parallel baselines that are similarly spaced. From these the GPIR system is deployed at right angles enabling a consistent pattern of data to be collected. This process is repeated at 90 degrees to the first to complete the grid. Using this method enables a more reliable and consistent set of results to be produced. In particularly difficult areas or where there is a high likelihood of non linear targets being present, it is not uncommon for a second orthogonal grid to be constructed at 45 degrees to the first. This improves the clarity of detection with respect to short runs, T connections etc. The intensity of the grid can also be altered by increasing or decreasing the number of reference points and the distance between the baselines used. The greater the number of data sets collected, the more comprehensive the survey, but the greater the time taken to complete it.

DATA INTERPRETATION

Accurate interpretation of radar data is a specialist skill and should not be attempted by those without the required training and experience. Depending on the level of accuracy demanded and the time frame in which the results are required, the data may either be interpreted during the collection process by a highly trained specialist, or collected by a technician and interpreted by a specialist at a later date. In some circumstances such as where a QA process is required it is often found prudent to do both.

In addition to the software packages supplied by the equipment manufacturers, several after market packages are also available.

ADVANTAGES & LIMITATIONS

In addition to the detection of services locatable by more traditional means, GPIR is unique in being able to detect non-conductive and non-metallic services just as easily. It can also assist traditional methods such as E/M location in those areas where the E/M field has been disturbed by the presence of High voltage power lines or, where no direct access point is available for the connection of a signal generator (Genny).

In areas with a low and consistent water level, it is possible to locate the presence of a water leak from a nearby supply pipe by identifying areas with an unexpectedly high concentration of water. This is possible because of the way a GPIR signal interacts with water. It is also possible to identify voids washed out by water leaks for the same reason.

The current level of technology does not permit the accurate identification of size, purpose, or material type of any changes located and areas with different surface coverings i.e., tarmac, grass, concrete etc. it is advisable to survey each surface separately, to prevent any deterioration of data clarity.

Areas of ground with high clay content and/or a high water table are unlikely to yield results as conclusive as those areas which are sandy or dry, or any combination between in varying degrees. Those areas where the services are constructed of materials with similar dielectric properties to the surrounding ground may also prove problematic. Areas with a high saline content are similarly problematic.

Although the material properties of the ground being surveyed can affect the overall quality of data obtained, particularly in relation to the depth from which a discernable radio signal can be received, the main limitation is the frequency of the transmission. The higher the frequency used; the shallower the depth that will be achieved. Transmit frequencies centred on 250 MHz for example will typically reach depths of around 3 metres. For those centred on 1 GHz, depths of 1 metre or less are more typical.

The frequency used also determines the minimum physical size of any material change that can be detected. This is referred to as the resolution value. It also affects the minimum spacing between adjacent material changes that can be detected. This is the discrimination value. Any material change or spacing between adjacent changes that is of a size greater than 50% of the wavelength of the transmitted frequency, tends to be detectable more often than not. Any change or space smaller than 25% is rarely detectable.

The difference between these guide limits is the grey area where the affects of poor ground conditions, technique or inexperience of the operator are most noticeable.

The following chart provides a useful guide:

Typical duration of signal pulse (Nanoseconds)	Centre transmit Frequency (MHz)	Approximate depth of penetration (Millimetres)	Approximate discernable target size (Millimetres)
1	1,500	300	30
1.5	1,000	1,000	100
3	500	2,000	200
6	250	3,000	300
15	100	5,000	500
30	50	10,000	1,000
75	20	20,000	2,000

These are typical values for unsaturated, naturally damp sandy soil. In practice the depth and discernable target size may be very different and in most cases shallower and smaller than above.

Where no supporting corroboration of target depth is available, it is recommended that a value equal to $\pm 10\%$ of the depth given is allowed for.

In order to mitigate the effects such limitations have on the quality and integrity of data, it is not unusual for a GPIR survey to exceed the area required by as much as 30%.

LEGISLATION & STANDARDS

In this context, GPIR systems are no different from other radio systems. The radiation emitted from GPIR systems has been carefully evaluated as part of the licensing process, and it is most unlikely that GPIR systems will interfere with other radio systems. Anyone operating a GPIR system should ensure that they are in compliance with all legislation relevant to the place of use before proceeding. In the UK for example, it is a requirement that all companies operating GPIR systems are licensed and keep a detailed "Log of Use". A similar requirement may be extended across the EU in the near future.

The recognised trade body for Ground Probing Radar users within Europe is the European GPR association. (EuroGPR). Further details of this organisation can be found at www.eurogpr.org.

The following are known standards and guidelines covering the use of GPR for the detection and location of buried services:

- EuroGPR policy statement on utilities detection and mapping.
- Association pour la qualité en Géophysique Appliquée non Pétrolière (AGAP), Guide d'adéquation des méthodes et techniques géophysiques aux applications concernant le sol et le sous-sol (guideline for the update of geophysical methods and techniques for soil and sub-soil applications), 1992.
- American Society of Testing and Materials (ASTM), Standard Guide for Using the Surface Ground Penetrating Radar Method for Subsurface Investigation, ASTM D 6432-99, 1999.
- American Society of Civil Engineers (ASCE) Standard Guideline for the Collection and Depiction of Subsurface Utility Data, CI/ASCE 38-02, 2002.
- Comitato Elettotecnico Italiano (CEI) Norma per l'impiego del radar per introspezione del suolo per prospezioni preliminari ad opere di posa di servizi ed infrastrutture sotterranee (Regulations for performing preliminary surveys with ground probing radar before laying underground utilities and infrastructures), CEI 306-08, 2004.
- Fondazione Politecnico di Milano, Sistemi geognostici per la mappatura dei sottoservizi (Mapping subsurface utility networks using geognostic investigations).
- The Survey Association (TSA) Underground Utilities Surveys; Detailed guidance notes for specifying a 'utility' survey, Draft 5 (To be published), 2009.

In addition to the standards and guidelines listed above, there is also a guideline concerning the use of GPR in sensitive radio environments. Originally issued by EuroGPR for the benefit of its members, this has now been adopted as a "Technical Report" by European Telecommunications Standards Institute (ETSI) and at the time of writing was under review for advancement to "European Guidance" status.

USE NEAR SENSITIVE RADIO ENVIRONMENTS

The modern world relies upon a wide variety of radio systems for many diverse activities. These normally operate successfully together because the compatibility issue for each has been carefully considered in terms of the context of the radio spectrum however occasionally due to exceptional situations interference may occur.

GPIR systems are typically designed to radiate low level signals into the ground or, for example, the concrete, brick or stone of a civil engineering or building structure. Under these circumstances there is little likelihood of high levels of signal escaping into the air. The most sensitive radio environments are those where the safety of life or similar function may be threatened if interference occurs to a radio system. Generally the areas of most concern are:

- Airfields
- Prisons
- Defence establishments
- Radio astronomy sites

When within 1.5 km of these and other sensitive locations, GPR/GPIR operators should be particularly aware of the potential for interference and may be required to lodge details of their equipment and frequencies to be used prior to being given clearance to proceed. In the UK the main radio astronomy sites near which restrictions apply are at Cambridge, Darnhall, Defford, Joddrel Bank, Knockin and Pickmere.

At all times, operators should ensure that they minimise the chances of stray signal radiation by:

- Only keeping the GPR system or its transmitter switched on when actually carrying out measurements.
- Whenever practical ensuring that the GPIR antenna is in contact with a layer of absorptive earth material of sufficient thickness and properties to absorb and dissipate the GPR's low level radiation.

A log should be kept of every occasion upon which GPR/GPIR is used. The log should include the map reference of the location; details of the equipment employed particularly frequency of transmission, the time and date of operation and the name of the operator.

REFERENCES

Farley, M. and Trow, S. (2003) Losses in Water Distribution Networks – A Practitioner's Guide to Assessment, Monitoring and Control. London: IWA Publishing. (http://www.iwapublishing.com/template.cfm?name=isbn1900222116).

Farley, M. and Hamilton, S. (2008) Technology and Equipment for Water Loss Management (to be published by IWA Publishing in 2008). http://www.iwapublishing.com/template.cfm?name=isbn1843390868

Conference proceedings, Leakage 2005 – IWA Specialized Conference, Halifax, Nova Scotia, Canada, 12–14 September 2005: www.leakage2005.com

Conference Proceedings, Global Leakage Summit, London, UK, 24–26 January 2007. London Business Conferences: www.global-leakage-summit.com

Marshallsay, D. and Bond, A. Conference proceedings, *Water Loss 2007*, 23–26 September 2007, Bucharest. Volume 3, pp. 818–826: http://www.waterloss2007.com

Thornton, J. (2002) *Water Loss Control Manual*, New York: McGraw-Hill, pp. 227–233.

Trust, J. and Allen, K. (2003) *Ground Penetrating Radar – Detecting Leakage in Water Networks*. Geo Connexions.

Kiss, G. Koncz, K. and Melinte, C. Conference proceedings, *Water Loss 2007*, 23–26 September 2007, Bucharest. Volume 3, pp. 622–631: http://www.waterloss2007.com

ETSI EN 302 066-1 "Electromagnetic compatibility and Radio spectrum Matters (ERM); Short Range Devices (SRD); Ground – and Wall-Probing Radar applications; Part 1: Technical characteristics and test methods.

ETSI EN 302 066-2 "Electromagnetic compatibility and Radio spectrum Matters (ERM); Short Range Devices (SRD); Ground – and Wall-Probing Radar applications; Part 2: Harmonized EN under article 3.2 of the R&TTE Directive.

ETSI EN 301 489-32 "Electromagnetic compatibility and Radio spectrum Matters (ERM); ElectroMagnetic Compatibility (EMC) standard for radio equipment and services; Part 32: Specific conditions for Ground and Wall Probing Radar applications.

EuroGPR Code of Practice for the Use of GPR Systems in Potentially Sensitive Radio Environments.

Chapter 5

Usage of GPIR devices: The WATERPIPE ground penetrating imaging radar – GPIR

Nikolaos Uzunoglu, Matthaios Bimpas, Yorgos Stratakos, Michail Sofras and Angelos Amditis

The present chapter provides the description of both versions of the implemented prototype GPIR systems, as well as some information for the leaking radar system that has been integrated with the WATERPIPE system. It gives a good overview of the main technical system characteristics, the main subcomponents of the GPIR and the leaking radar sensor.

METHODOLOGY

In designing the prototype system, a priori technology selections has been at the proposal level. These fundamental choices are:

- Use of Ultra Wideband – short pulse transmission to illuminate the earth medium.
- Measure the reflected waves on a surface of sufficient horizontal area to achieve a good spatial resolution in three dimensions.
- Utilize mechanical and electronic scanning to measure and acquire the reflected-scattering signals.
- The processing of measured signals should be done on site and the presentation of images should be achieved in three dimensions on "real time".
- In order to explore wider areas the system should be moved along a linear or curve path.

The definition of the Fundamental Operational Specifications (FOS) of the system is of paramount importance issue, since their selection should comply with two requirements:

- Achieve the required imaging and operational characteristics of the GPIR on detecting water pipes faults and their nature.
- Not violate existing technology capabilities and of course physical limitations imposed by E/M theory.

Taking into account the above two conditions and the nature of the water pipe faults the following FOS are selected:

(1) Maximum Imaging Depth: 2 meters from the earth surface.
(2) Spatial-imaging resolution in three dimensions: better than 5 cm.
(3) Acquisition time of underground image of 1×1 square meter (m^2) in less than 10 seconds.
(4) The radar should be capable to be mounted on a truck to carry continuous measurements along a path.
(5) Power consumption requirements: less than 500 Watt.

©2010 IWA Publishing. *Integrated High Resolution imaging Radar and Decision Support System for the rehabilitation of WATER PIPElines*. Edited by Matthaios Bimpas, Angelos Amditis and Nikolaos Uzunoglu. ISBN: 978184333719. Published by IWA Publishing, London, UK.

(6) The operation of the GPIR system should be fully automatic without any intervention of the operator and only to be on the level of supervision to take action in case something to go wrong because of the prototype nature of the system.

(7) The GPIR should be capable of surveying 1km in a time of less than 2.8 hours.

Based on the above specifications the system specifications were defined taking into account the available E/M and signal processing technologies.

IMPLEMENTATION OF THE GROUND PENETRATING IMAGING RADAR – GPIR

Operation principle

The operation of the High Resolution GPIR is based on the following principles:

- A short duration (nanosecond) pulsed signal is generated periodically by the short pulse transmitter.
  - The Pulse Repetition Frequency (PRF) is several MHz and could reach 10MHz.
  - The high PRF is useful to achieve the generation of three dimensional underground environment images in minimum time.
  - The output pulse is driven to a wide band transmitting antenna emitting the pulsed radiation towards the earth.
  - The energy content of the pulsed is peaked in the spectral region 500–1500 MHz depending on the selected pulse duration.
  - Three distinct pulse duration were selected in the prototype to be developed which are 2, 1 and 0.5 ns.
  - The limited size of the TEM horn type wide band antenna results into a high pass filter performance and then the radiated energy has zero baseband spectral components.
- The Short Pulse Generator outputs amplitude of approx. 20 $V_{p\text{-}p}$ (peak to peak) and drives to fixed position wide band antenna.
  - The transmitting antenna is of TEM horn type consisting of two flared trapezoid shape metallic surfaces as described below.
  - The distance of the TEM horn antenna aperture is at 20–40 cm from the earth surface.
  - The exact distance maybe modified during the experiments.
- The radiation of the pulsed signals results into illumination of the ground environment with the penetration of the wide band signals. The E/M properties of the earth medium being of a dielectric medium with losses results into strong dispersion phenomena. In general terms the earth medium being inhomogeneous and because of the existence of artificial materials such as water pipes which is of our interest. Therefore the response of the underground medium has a highly complicated nature in terms of E/M properties. In any the system to be developed aims to image the penetrating E/M field distributions.
- The reflected-scattered fields form the underground environment is measured and recorded on a surface of $1 \times 1\text{m}^2$ parallel to the earth medium. The measurement of field on this aperture is carried out with a density of 0.1×0.1 cm. In the following a detailed description of the scanning antenna incorporating both an electronic and mechanical scanning is described in detail.
  - The selection of the receiver antenna parameters are determined by the requirement of achieving underground three dimensional imaging resolutions of 5 cm in all dimensions.
  - The use of hybrid scanning technique results at any given instant into a single output, thus requiring only a single receiver channel which provides an effective multiplexing.
  - The received signal is then filtered and amplified to through a linear channel. This analog signal conditioning prepares the signal to drive the High Speed Analog to Digital Converter (A/D-C) operating at 1.5 G-Sample/second rate.
  - The sampling is carried out with an 8 bit resolution and the measurement voltage value is stored at a buffer memory. In each measurement point averaging of recorded signals is carried out to improve the signal to noise ratio of the received signals.
  - The number of averaged pulses is of the order of 8–16 or even to 64 if this is required. The use of high PRF allows the averaging large number of pulses. The A/D-C circuit is triggered by the same master clock triggering the short pulse transmitter.
- The use of UWB signals provides extensive information for the imaging of the underground target medium.
  - At the end of each receiver scanning antenna, up to 400 reflected-scattered signal waveforms are available to the processing unit. The reflected scattered waveforms are produced by acquiring, averaging and storing the real and imaginary parts of signals peaked by the 10 probe antennas placed along a linear line which in turn also scans the aperture along a vertical line. In the following section the operation principle of the scanning receiver antenna is explained in detail.

- The reconstruction algorithms of the underground induced field and medium complex dielectric constant (two real numbers for each frequency spectral component) are obtained using a time domain version of the Method of Auxiliary Sources (MAS) which has been developed in the past.
- Two alternative versions of the reconstruction algorithm are being experimented to observe the performance and selected the most accurate one. Essentially the reconstruction of the complex electric field of the underground medium with the embedded water supply pipe poses an "Inverse Scattering" problem whose solution leads into imaging of the medium under measurement.

The prototype GPIR

The first prototype WATERPIPE GPIR system has been implemented and constructed since September 2007 and since then a series of measurements have been carried out, in order to validate its capability to perform the desired imaging and detect the water leaks. Some photos of this prototype are given in the Figure 5.1.

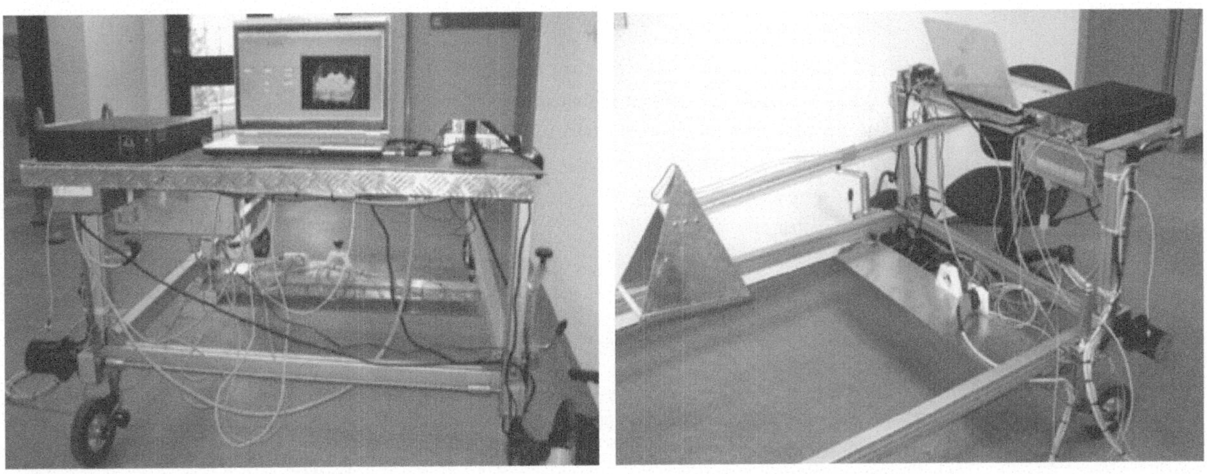

Figure 5.1 Photos of the GPIR System. (Left photo) The User Interface (UI) can be discriminated, (Right Photo) The procedure of measurements is presented.

After the first tests of the prototype GPIR, it was considered essential to redesign the equipment in order to perform more efficiently and robust. At the following Figure 5.2, some photos of the refined prototype are presented. The main advantages of this refined version of the prototype compared to the previous one are:

- The size of the system is rather smaller, that makes it more practical in use. The whole system is also lighter in weight.
- The use of one main receive unit Low Noise Amplifier (LNA plus filtering units) makes the performance rather robust, not needing any special calibration before using it.
- The design of the transmit antennae as well as their placement are theoretically expected to be better than the ones used in the previous system.

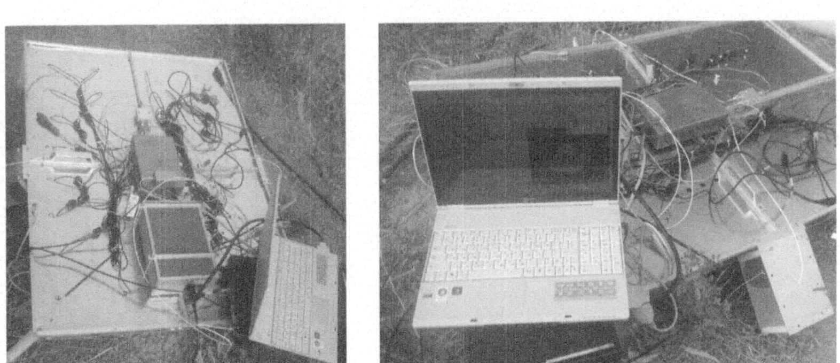

Figure 5.2 Photo of the 2nd version of the prototype system.

BASIC HARDWARE SYSTEM COMPONENTS FOR THE FIRST GPIR

The electronics of the system can be divided into 3 parts:

- RF generation
  - Pulser (RF & trigger signal)
  - Antennas (Transmit and Receive)
- Acquisition
  - Single Pole to N Throw Switches (mechanical)
  - Acquisition of Incoming RF signal (sampling and storing)
- Control
  - Control of the moving Scanner Element (consisting of 10 antennas)
  - Control of all sub-modules
  - Master Synchronization of the System

Transmitter unit

An UWB signal generator is utilized to produce sub-nanosecond pulse (500–3000ps). The peak pulse amplitude is of the order of ±10 Volt. This pulsed signal is conveyed to the transmitting antenna which is a Transverse Electromagnetic (TEM) Flared Horn antenna as shown also below. The transmitting antenna (see Figure 5.3) is fixed contrary to the receiving antenna which is of scanning type. In order to decrease the total sampling time of the reflected signals from the underground, the pulse repetition frequency (rate) should be up to 10MHz. The use of TEM horn guarantees wide band matching of the antenna of pulse generator.

Electrical specifications, as were initially set are presented in Table 5.1:

Table 5.1 Specifications set for the GPIR transmitter.

Specification	Value	Units
UWB Spectrum Centre Frequency	800	MHz
Total Spectral Bandwidth	300–2000	MHz
Pulse amplitude (pk-pk)	5–20	Volts
Pulse duration (min/max)	0.5–3	nsec
Repetition Frequency	10	MHz

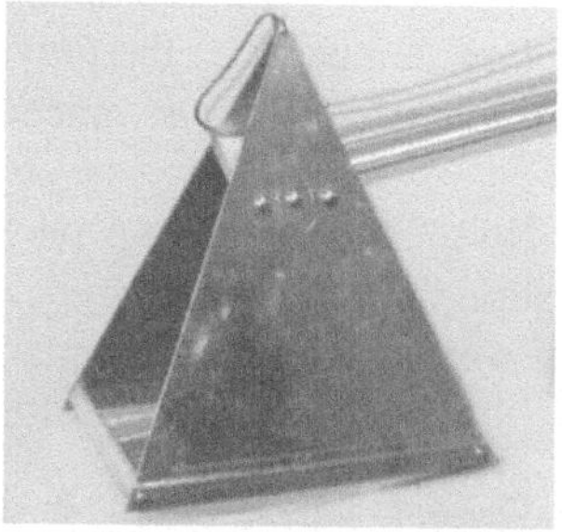

Figure 5.3 TEM Horn transmitting antenna.

Receiver front end

Signal conditioning of the scanning antenna should be carried out using a Wide Band Low Noise Amplifier (0.4–2GHz) and an Anti-Aliasing Filter. The Low Noise Amplifier gives a maximum gain of 40dB gain. The received echo pulse signals reflected by the ground structure are amplified up to a level that is sufficient to convert into digital signals. Direct Analog to Digital (A/D) conversion to achieve robust signal detection of echo signals. The following specifications – as provided in Table 5.2 – are set up concerning of the receiver front end:

Table 5.2 Receiver specifications.

Measure	Value	Units
Gain	38±1	dB
Frequency	500–1000	MHz
Noise Figure	4.5	dB
VSWR $_{IN,\ OUT}$	1.5:1	–
Input 1db Compression Point	–30	dBm
Isolation In Between Channels (Min)	45	dB
Power Consumption	+5V/100mA	–
Control	USB 2.0 control	–

The Scanning Receiver Antenna, as depicted in Figure 5.4, is the most critical subunit of the GPIR. The antenna scans mechanically parallel to earth surface above the earth surface approximately 20–30 cm.

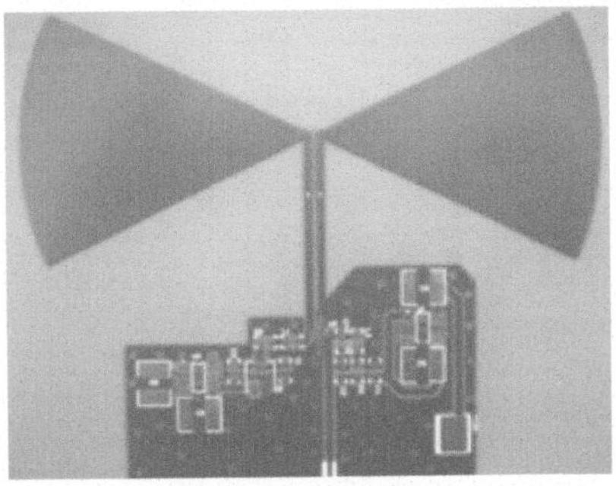

Figure 5.4 Photo of the antenna.

The scanning is of raster type that provides scanning of 1×1 m^2 horizontal area in the Cartesian coordinate system as shown in Figure 5.5. A sliding bar carrying out 10 wide band active short size (with respect the average wavelength = 35 cm) is used to scan the aperture of 1×1 m^2.

The bar motion is linear with a speed of 0.1–0.2 m/sec.

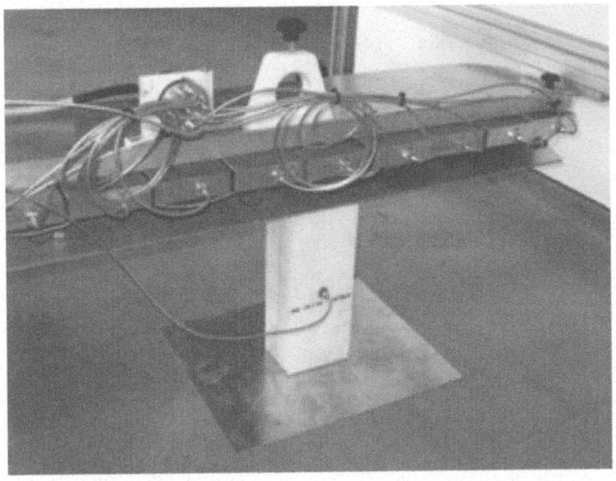

Figure 5.5 Photo of the receiver.

The switching among the 10 shaped antennas (please see Figure 5.6) is achieved by using pin-diode switches and activating at a given time 1 out of 10. The use of PRF as high as 10MHz provides the opportunity to accomplish the acquisition of 10 antenna signals with 64 sequential pulse averaging in a time of $64 \times 10 \times 0.1$ μs that equals to 64 μs. Therefore the dipole bar motion is irrelevant with respect the signal acquisition time. This also means that much higher scanning speeds could be achieved in recording the aperture fields reaching to an acquisition time of 1–2 seconds. This fact is highly important since this means that the radar could operate in a slowly moving car and increase drastically the overall inspection time of pipes with the GPIR under design. The geometry of the whole implemented system is provided in Figure 5.7.

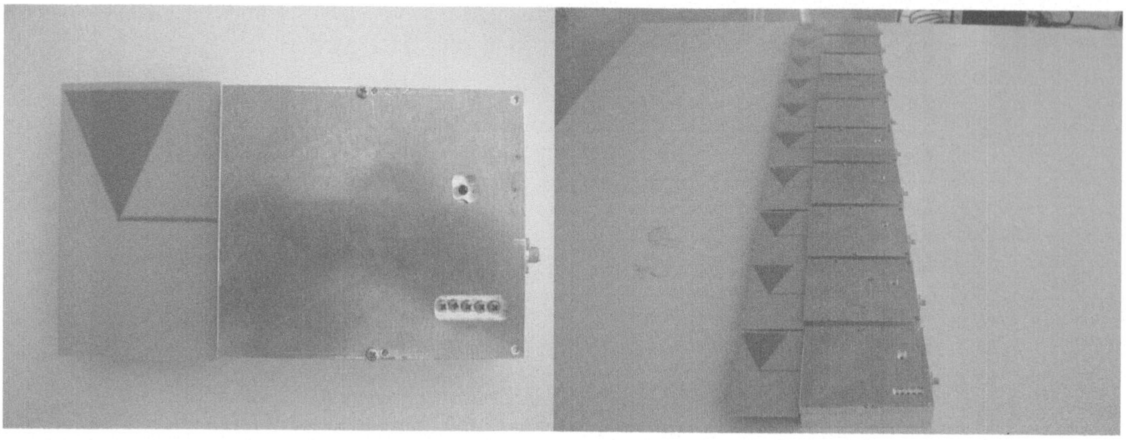

Figure 5.6 Photo of the antenna/receiver and the whole receiving unit by putting 10 separate receiving modules in a row.

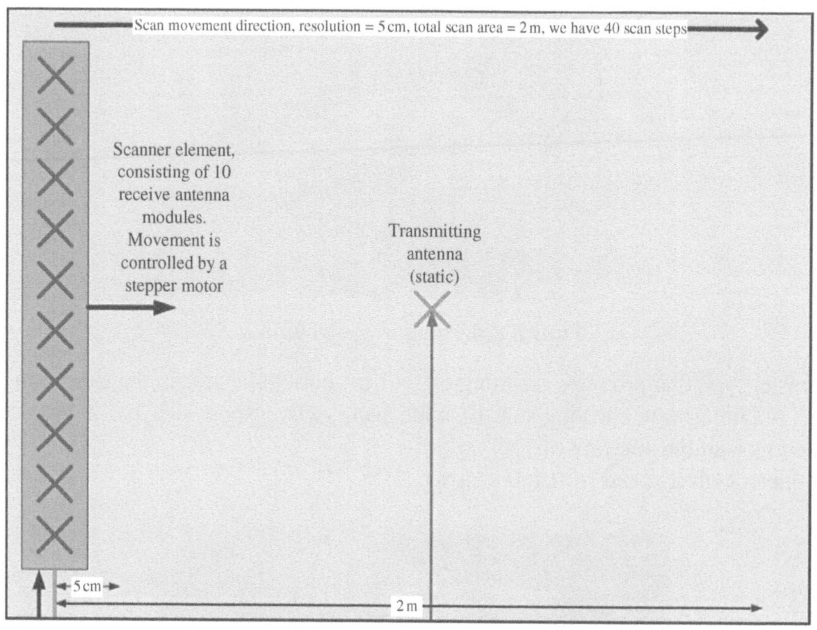

Figure 5.7 Receiver antenna block & scanning geometry.

Signal digitizer, computer and image display

Designing and building a system suitable for the WATERPIPE needs demands several technical aspects to be taken into consideration, since the final demonstrator is not a pure lab based experiment. Series of outdoor measurements in real conditions and at various terrains have to be taken during the whole procedure of design and implementation. So it is crucial for the further progress of the project t carefully select or even built a digital system, robust, fast and expandable without forgetting in any procedure of design the need for portability and compactness. So the issues of portability and thus the potential lack of strong and efficient power sources at the fields of Testing lead our entire research and design to implement clever architectures that can give us a low power consumption with power saving options digital system, yet able to meetall the present and future needs of the project.

This digital system provides an interface with the RF part of the system, and after sampling all the data provided to that interface, it basically performs digital signal processing of the incoming streams and using innovative algorithms it finally gives us the three dimensional representation of the underlying pipe infrastructure at the scanned area of interest.

The heart of the system is a super fast data acquisition and control sub-system, combined with an x86 processor based platform, a laptop in our case, which is responsible for the post processing and analysis in order to implement the 3d reconstruction of the pipes. So logically the digital system is divided in the two parts.

The first part, the data acquisition and control module, has to be able to perform sampling at rates of no less than 1GSps, properly buffering and transferring the valid data to the second part system. The desired resolution of 8 bits must be met in order to achieve the necessary Signal to Noise Ratio (SNR) for the proper function of the algorithms on the next step of processing.

The second part of the system obtains the sampled data from the first part, and performs high speed post-processing in order to calculate the required visual 3D output. In order to achieve these tasks we use a Laptop with the proper configuration, that offers us portability and an environment that everyone is familiar with. So by running the proper application software on the Laptop, we achieve communication with the acquisition and control part of the system, and the input data can be processed with our software achieving the required results on the screen. The final design of the complete digital implemented subsystem is depicted on Figure 5.8.

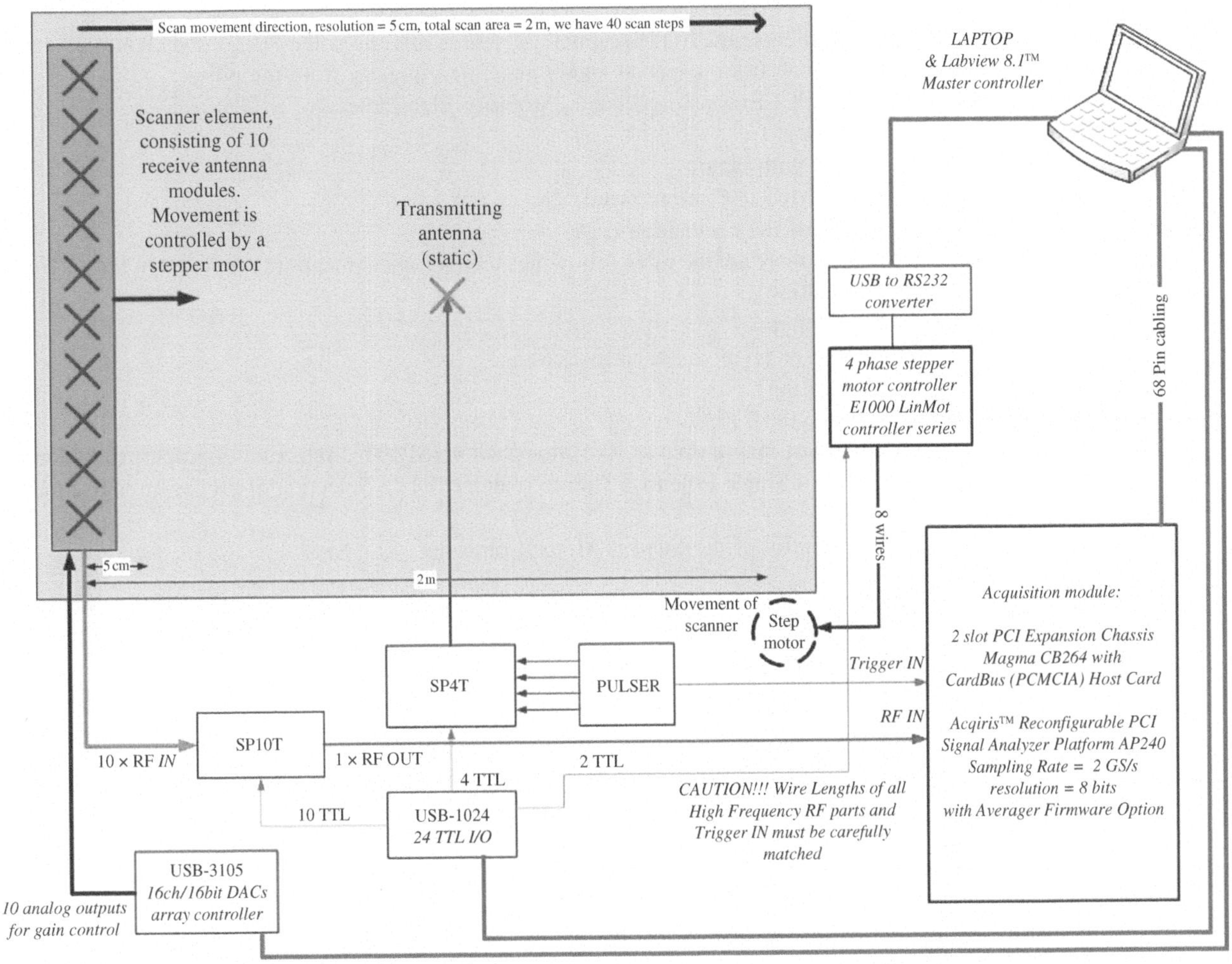

Figure 5.8 Schematic diagram of the implemented system.

The main technological components used for the acquisition procedure (Figure 5.9) are the followings:

- AP 240 Acqiris™ PCI Signal Analyzer Platform
  - 2 channels × 1Gsps, 8bits resolution
  - 1 channel × 2 Gsps, 8bits resolution (interleaved)
  - External Trigger input
- Averager Firmware
  - Synchronous, dual-channel real-time sampling and averaging at up to 2GSps
  - 12 bit dynamic range achieved with dithering in SuperRes mode
  - Round robin and standard segment averaging

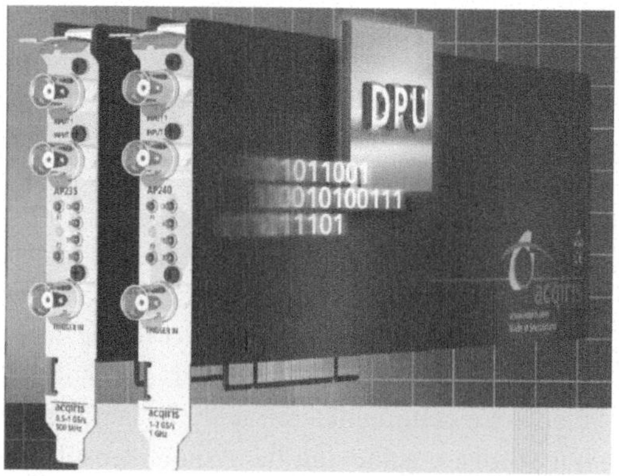

Figure 5.9 Data Acquisition Card AP 240 Acqiris® PCI.

The RF Signal from the selected "read" antenna of the scanner element is driven via the SP10T switch and guided to the RF in of the AP240. Triggering is done from a special signal generated directly from the pulser.

AP240 is hosted in an external Magma PCI Expansion Chassis, Mobility Electronics

- 2 Slot PCI Expansion, CB264
- PCMCIA card for connection with laptop
- 68 pin cable from rack to PCMCIA (Cardbus) card.

Averaged data are stored in memory of the acquisition card.

As far as the control unit is concerned, there are the main followings technological means (as depicted in Figure 5.10):

- USB-3105, Measurement Computing
  - 16 channels, 16-bit analog output
    - 10 used for gain control of receiving antenna modules
- USB-1024, Measurement Computing
  - 24 Transistor-Transistor Logic (TTL) I/Os
    - 10 Outputs used for control of Single Pole to 10 Throw Switch (SP10T). Selection of proper read antenna.
    - 4 Outputs used for control of Single Pole to 4 Throw Switch (SP4T). Selection of proper timing sequence of the pulser.
    - 2 Outputs used for state control of the Stepper Motor Controller.

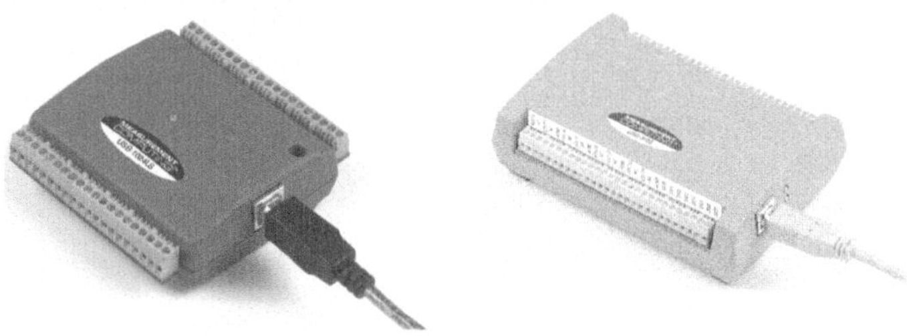

Figure 5.10 Left: TTL I/Os controller (USB-1024LS), Right: Gain controller for receiver modules (USB-3105).

As for the Stepper Motor Control (please see Figure 5.11), we used:

- LinMot E1000 and PowerSupply, LinMot
  - 4 phase stepper motor controller
  - RS 232 interface for communication with PC
  - Precise Motion Control Curve
- USB to RS232 converter, generic
  - For interconnection with the Laptop

Finally, concerning the master unit:

- Laptop, LG electronics
  - Intel Core 2 Duo, 1.8GHz
  - 2GB RAM
  - 120 GB HDD

  - 4 USB
  - Card Bus interface
- LabView 8.1, National Instruments
  - With drivers for all the interconnected devices
  - Program developed in graphical language G, implementing Virtual Instruments and control algorithms – sequences
  - Target is a final stand-alone application.

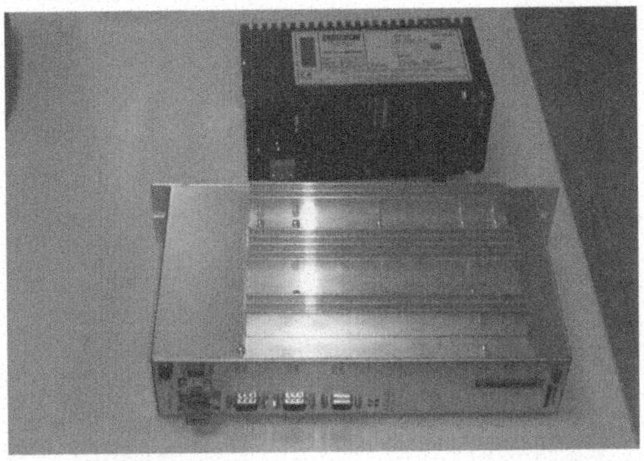

Figure 5.11 Stepper Motor Controller LinMot E1000 & Power Supply.

MODIFICATIONS IN THE UPDATED VERSION OF THE SYSTEM

The second version of the GPIR uses fixed antenna system avoiding any mechanical motion in collecting the scattered fields. The collection of data is carried out through electronic scanning of 36 antennas and collecting along the time axis the reflected signals from the ground medium.

In Figure 5.12 the geometry of the $4 \times 8 = 32$ array antenna and the transmit antenna are shown. It should be emphasised that both the antennas for the transmitter and the receiver are identical.

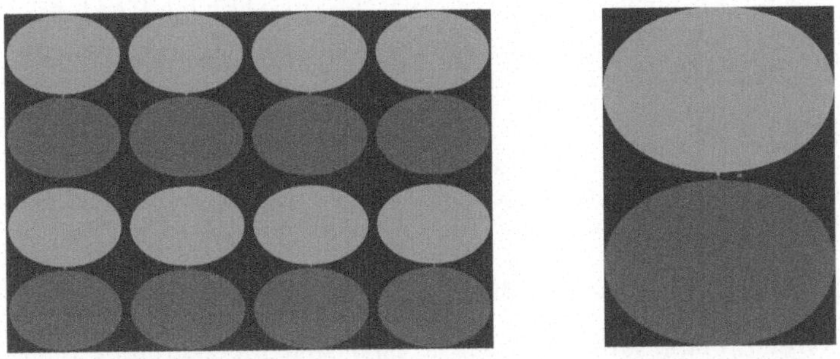

Figure 5.12 Schematic of the antennas employed in the Receiver Unit- only 16 of receivers are presented, there is one more panel with 16 receive antennas (left) and the transmitting antenna- only one of them (right).

At Figure 5.13, the antenna gain pattern along with the S11 parameters are presented for the single antenna element.

At Table 5.3, the main characteristics for the antenna are provided.

The signals derived from each antenna element are driven to an analogue switch which is used as a multiplexer. In order to have a single measurement channel a cluster of analogue switches are used to connect each time an antenna to data acquisition unit. This means that each one of the 32 antennas is "visited" consequently. The trigger to start to carry out measurements is obtained from a short pulse signal generator.

The UWB character of the system makes the receiving channels to be open for any signal incident from other sources such as Television or Radio Transmitter signals. In order to suppress these interfering signals the averaging procedure is used and in this context on each antenna data is measured is repeated and an averaging is carried out.

The externally incident fields being incoherent along the time axis are wash out after averaging the measured data while the coherent signals refereed to the same trigger signals are enhanced.

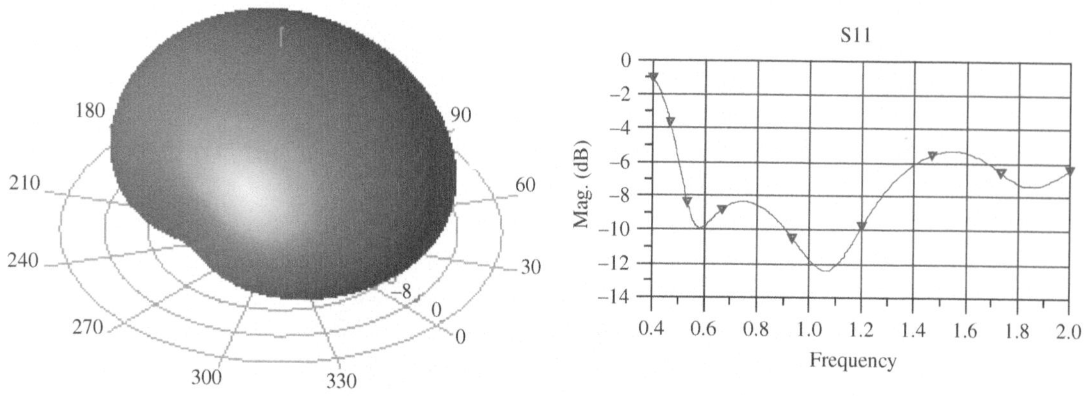

Figure 5.13 Antenna 3D Gain pattern (left) and S11 parameters (right) for the single element (both receiver and transmitter).

Table 5.3 Antenna characteristics for the single element.

Power radiated (Watts)	0.00226898	
Effective angle (Steradians)	2.54251	
Directivity(dB)	6.93947	
Gain (dB)	6.92496	
Maximum intensity (Watts/Steradian)	0.000892417	
Angle of U Max (theta, phi)	28	359
E(theta) max (mag,phase)	0.0113131	−101.565
E(phi) max (mag,phase)	0.819923	18.6687
E(x) max (mag,phase)	0.0126724	−24.2467
E(y) max (mag,phase)	0.819886	18.6792
E(z) max (mag,phase)	0.00531116	78.4352

Theoretically if we use an averaging of $N = 100$ signals a $20\log N = 40$dB suppression of interfering signals is achieved. The consequently collected signals from the antennas are stored on the RAM of a lap top computer and after storing the completion of signals', a software program developed for this purpose is utilized to develop 3-D images.

SOFTWARE DESCRIPTION

The developed software consists of three main components.

1. Software for acquisition and control

The first one is the acquisition and control software, which is responsible for controlling all the system hardware, i.e., the receive antenna switches, the transmitting waveform switch, the stepper motor controller and the gain amplifiers. It is also responsible for the control and acquisition of the data using the acquisition card, as well as for the synchronization between transmitter, receiver units and acquisition.

The software for this part was implemented entirely in LABVIEW©, and presented in the following paragraphs.

Computer and image display

The Laptop is the master mind of the system, connecting and controlling the Data Acquisition System, the Selector Switches SP10T and SP4T via a USB to TTL I/O interface, the Gain Control of the Amplifiers using a USB to 16 channels ADCs of 16 bits resolution interface. The Laptop also runs the software for the Image Reconstruction.

A Virtual Instrument (Program) was developed in LabView™ that executes the following descriptive algorithm:

(1) Selection of the proper timing sequence from the pulser, calibration of the Amplifiers' gain, Calibration of the Motor Controller.
(2) Movement of the scanner element to the right position, i.e., scan line = 0 for the first scanning line.
(3) Selection of the right antenna on the scanning element, i.e., scan antenna = 0 for the first antenna.
(4) Triggering, Acquisition and Averaging of 10 i.e., sampling snippets, 16 samples per snippet.
(5) Transferring the averaged snippet to the Laptop and storing into the right place in a matrix.

(6) Repeat step 3 until a full scan line is sampled.
(7) Move the scanner element to the next position. Advance the stepper motor the proper steps until the element has advanced 10 centimeters in total.
(8) Repeat step 2, until 2 meters are scanned, giving us a total of 20 scan lines, with 10 snippets per line, of 16 samples of 8 bits per snippet.
(9) Export Data to DSP and Image Processing Module with data.
(10) End program or repeat from the beginning as desired.

Signal digitizer

The acquisition is performed by a PCI Signal Analyzer Platform card from Acqiris and specifically the AP240 that provides us with two channels of 1GSps each at 8bits resolution and in interleaved mode one channel of 2 Gsps at 8bit, combined with a special dithering mode that increases the dynamic range up to 12 bits. It also provides in hardware round robin and standard segment averaging which is ideal for our case.

The RF Signal from the selected "read" antenna of the scanner element is driven via the SP10T switch and guided to the RF in of the AP240, while triggering is done via the external trigger of the card that is connected to a special signal generated directly from the pulser. Extreme caution was taken in the wire lengths of triggering in respect to the whole signal path from generation to receiving. The averaged data is stored in a desired Matrix, in the memory module of the card, and then read at a much slower rate from the Laptop.

Choosing a Laptop for the portability, introduced further problems, since the AP240 is a PCI card and in terms of connectivity, we had to use an External PCI Expansion Chassis that could host the AP240 and provide the necessary bandwidth for transmitting the data to the Laptop. The Magma PCI Expansion Chassis from Mobility Electronics proved to be the solution to our problem. It provides 2 PCI Slots and interconnection with the laptop through a PCMCIA (Cardbus) Card, drivers and sufficient data transfer rate.

2. Software for image reconstruction

The second component of the developed software is the implementation of the reconstruction algorithms.

A novel reconstruction method has been developed, which reconstructs the 3D electric field and polarization currents in the area under the receiver block. The method results in a spatio-temporal convolution of the two-dimensional measurement data grid E0(x,y) with a four-dimensional reconstruction kernel U:

$$\vec{E}(x,y,z,t) = \frac{1}{8\pi^3}\int_0^t dt' \int_{-\infty}^{\infty} dx' \int_{-\infty}^{\infty} dy' U(x-x', y-y', z, t-t') \cdot \vec{E}_0(x', y', z' = 0, t')$$

where $U_\omega(\rho, z, \omega) = 2\pi \int_0^\infty dk_S k_S J_0(k_S \rho) e^{-j\sqrt{k_b^2 - k_S^2}\cdot z}$ written in polar coordinates, where

$$\rho = \sqrt{(x-x')^2 + (y-y')^2}$$

Software implementation consists of two main components. The first one calculates the kernel U using numerical integration, and is implemented in MATLAB™. The reconstruction kernel is calculation-expensive, and for a $1\times1\times5$ m^3 grid with a resolution of $10\times10\times10$ cm^3, it takes about 5 days on a dual core Pentium processor. However, the kernel is calculated once, and stored into memory. Thus, there are no speed requirements, and the code is optimized for convergence in the numerical integrations, and not for speed. A second similar kernel with second order Bessel functions is derived from the mathematical analysis for the polarization currents induced in the medium. This has been also implemented in MATLAB™ using similar code.

The second component of the reconstruction software retrieves and processes measured data, and subsequently implements spatio-temporal convolution of the measurements with the reconstruction kernel. This code briefly implements the following routines:

- reads the recorded waveforms from the acquisition card and removes DC component
- aligns successive acquired segments to remove triggering errors by using cross-correlation and averages them
- performs Fast Fourier Transform (FFT)
- performs two-dimensional spatial convolution with the reconstruction kernel stored in memory
- loops the previous operation for various depths

The code is optimized for speed, and currently takes about 20 s for a $1 \times 1 \times 5$ m^3 imaging region. The final version of the algorithm is exported to C, using the Matlab compiler (mcc). A wrapper C program calls the C version of the algorithm, and generates a dynamic link library (DLL).

There are four advantages of this approach:

(a) The development and testing time for the algorithm is short, since it is implemented in a high-level development platform such as Matlab.
(b) The final library executes fast, since it is C code and is executed without invoking the whole Matlab platform, but only the required subset of Matlab scripts.
(c) The library can be loaded into Labview, and get interfaced with the acquisition and control processes.
(d) The visualization and rendering part of the project can be incorporated into the Matlab code.

3. Software for visualization and rendering

The whole system has to be portable, and, thus, acquisition, control, image reconstruction and visualization runs on a single laptop PC. The hardware interface for control and acquisition of the system is implemented in Labview. Although Labview itself contains several graphics and visualization components, it is preferable being able to use dedicated software, which can take advantage of the capabilities of modern graphics cards. OpenGL is the industry standard library used for low-level 3-D rendering, and is implemented by all graphics card vendors, providing hardware acceleration and, thus, taking the load off the CPU. The Visualization ToolKit (VTK) is an open source, freely available object-oriented C++ class library software for 3D computer graphics and visualization, which contains many useful algorithms, importers, exporters and renderers. VTK has a huge set of classes, but is relatively easy to code. With very few calls, one can create an interactive and fast viewer/renderer, while properties for rendering primitives, lights and cameras are taken care of. VTK is loaded in MATLAB using the MEX interface. A basic vtk- MATLAB interface has already been developed. This software has been released as a MATLAB toolbox when completed. Note that this is only an optional add-on software component. Currently the system renders the reconstructed fields either in MATLAB or VTK.

LEAKING SYSTEM: A COMPLIMENTARY RADAR SENSOR EMPLOYED TO DETECT AND LOCATE WATER LEAKS

The WATERPIPE system as developed, it can not perform as efficient as to offer information about the existence of water leaks. Thus, it was decided by the beneficiaries to make use of another complimentary radar system that is capable of locating and detecting the water leaks. ICCS has worked on a previous research project so called LEAKING (co funded under the 5th EC FW); the WATERPIPE consortium considered that a re-design and integration of both WATERPIPE and LEAKING systems will lead to better outcomes and maybe this integrated system will prove to be a useful tool missing from the water utilities to make a mapping of the pipelines and at the same time search for leakages that the other conventional methods can not find.

Description of the system

Based on the previous gathered experience emanating from the implementation and performance evaluation of a Radar System for Detection of trapped humans under building ruins, the operational frequency of the radar system has been selected to be 2.45 GHz (Bimpas, 2004). This frequency combines a relatively good penetration depth with a high spatial resolution.

The block diagram of the implemented system is given in Figure 5.14.

A microwave Continuous Wave (CW) signal, generated by a very stable Voltage Controlled Oscillator (VCO), is filtered, amplified up to the desired power level by a Power Amplifier (PA) and transmitted by an antenna. The VCO is also used as a local oscillator in the receiver (homodyne topology), by making use of a Wilkinson divider (3 dB). At the receiver section, a microwave RF amplifier following a very sensitive LNA and a microwave band-pass hairpin filter brings the signal to a power level that can be processed by the mixer. Due to the extremely narrowband nature of the signal, equally narrowband filters were designed and constructed. Such filters are implemented in the baseband unit, for frequencies in which they are realizable. The digital processing enhances the performance of the filters. The mixer output is directed into the baseband unit. The latter is consisted of active, 4-pole, anti-aliasing low-pass filters with a cut-off frequency of 5 Hz, amplifiers and the A/D converter. The A/D converter used was ADC42 of Pico Logic. Its main characteristics are given in Table 5.4.

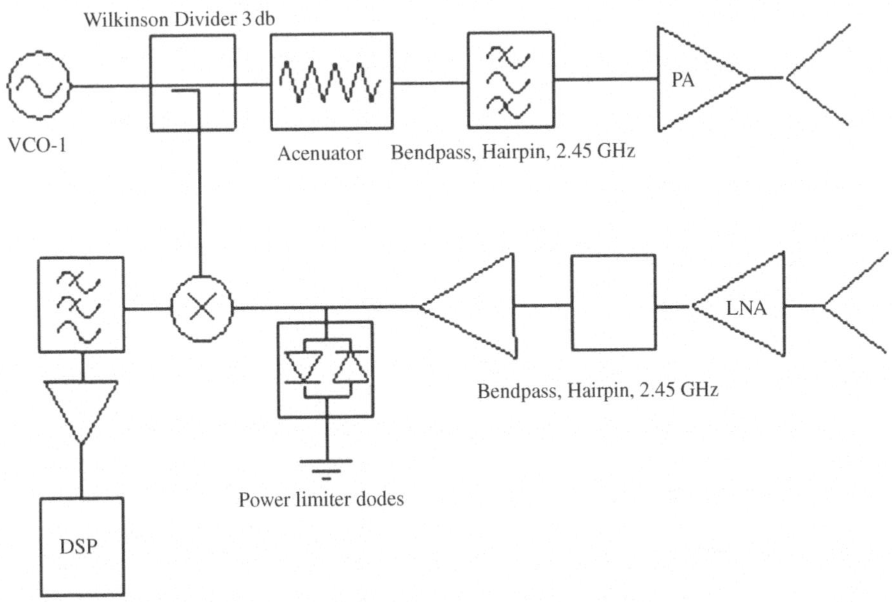

Figure 5.14 Schematic block diagram of the constructed Continuous Wave (CW) radar.

Table 5.4 Characteristics of the A/D converter.

Resolution	12 bits
Number of input channels	1
Input voltage range	−5 up to +5 Volt
Maximum sampling rate	15 k samples/sec
Typical error	±1% at 25° C
Input resistance	1 MΩ

The analogue signal coming out of the homodyne receiver is sampled with a data rate 300 Samples/sec, with an accuracy of 12 bits, following an analogue filtering with an equivalent bandwidth of 10 Hz. Then the stored digital voltage values are used to compute the FFT of the recorded signals. The Input voltage range of the A/D Converter is −5 Volt up to +5 Volts and its resolution is 12 bits, as shown in Table 5.4. Both spectral and time domain features of measured signals are observed to assess the existence of leaks and the type of them.

It was preferred to make use of a prototype microstrip rectangular patch antennae (with a toothbrush-shaped slotted parasitic patches and slotted ground for enhanced gain, as well as a dielectric cover for protection from ground surface roughness) for both the receiver and the transmitter (Bimpas and Uzunoglu, 2005). A photo of this prototype antenna used throughout all the testing procedure is presented in the following Figure 5.15. It is essential to emphasize on the optimised antenna's radiation pattern, in the great majority of the materials (concrete, brick walls, various types of soil), in which the E/M wave might propagate in real conditions.

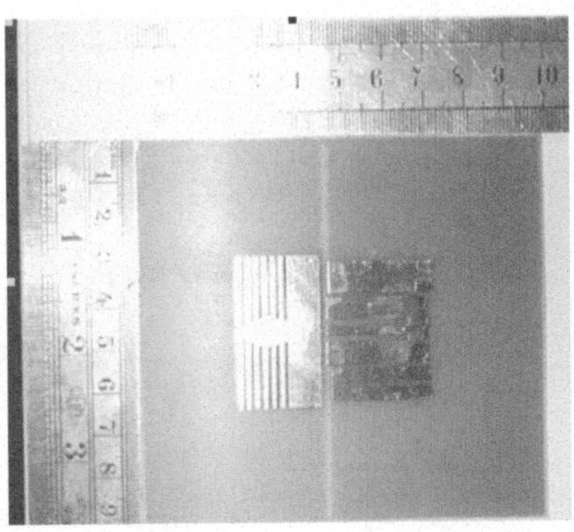

Figure 5.15 Photo of the prototype microstrip antenna. The dielectric cover has been removed so that the antenna construction is visible.

The main characteristics of this radar are presented in Table 5.5.

Table 5.5 Characteristics of the 2.45 GHz radar system.

Operation Principle	Continuous Wave @ 2.45GHz
Transmit power	2 Watt
Min Received signal	−110 dBm
Receiver Noise Figure	2 dB
Voltage Supply	+12 Volts
Power Consumptions	<5 Watts
Measurement Rate	1.5 meter (in the worst case)

Detection criteria for the water leaks

In order to assess the performance of the Water Leaking Detector Radar System extensive tests were carried out at appropriate test sites, in United Kingdom and Germany, with characteristics as similar as possible to the real water-leaking conditions. Moreover, this radar system was tested in real conditions in comparison to the prevailing methods for water leakages detection. In order to simplify the measurement procedure, all of the tested systems were placed on a wooden trolley, having appropriate holes on its lower surface for adjusting the radar system antennae. It was constructed in the National Technical University of Athens (NTUA) for this purpose and it is presented in the following Figure 5.16.

Figure 5.16 Photo of the trolley with all the equipment during the measurements procedure.

The most distinguishing feature of detecting water motion is the prevailing low frequency components in the range of 0.05 to 0.25 Hz, persisting across measurements. On the other hand, an irregular spectrum covering the entire bandwidth is observed in the absence of the water motion or in the case that the attenuation due to the underground material or the soil condition is high enough.

In Figure 5.17 a screenshot of a measurement taken on the leak point is presented. A relatively powerful (approximately 2V p-p) sinusoidal signal waveform in time domain and the concentration of frequency spectrum steadily below 2 Hz in the frequency domain clearly indicate the existence of a water leak in the specific test point.

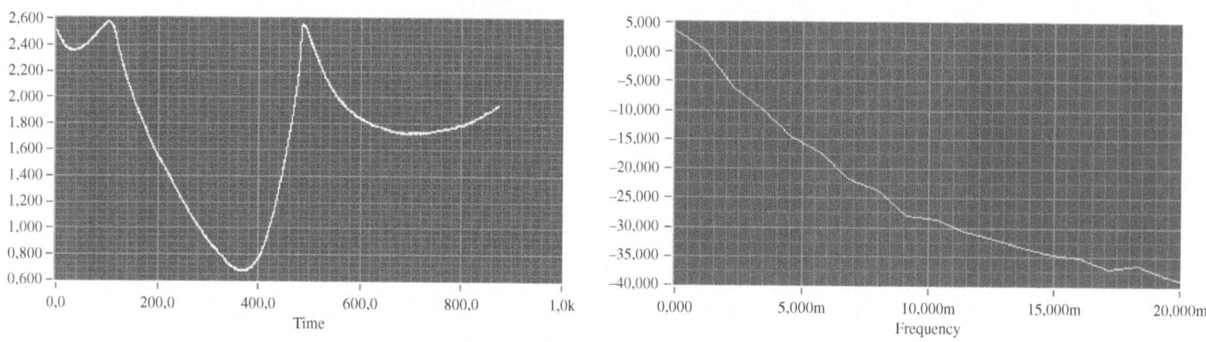

Figure 5.17 CW radar measurements on leak point: 2V p-p signal (sec – V) and spectrum (Hz – dB).

In Figure 5.18 there is a screenshot taken away from the leak. It is obvious that the signal is very weak (approximately 0.1V p-p) and noisy in time domain. Additionally, an irregular spectrum covering the entire bandwidth is observed in this case (the water motion is away from the radar sensor) in frequency domain. However it should be noted that the results might be the same if the attenuation due to the mass of material media (e.g. soil, cement etc.) is too high or in case that the water motion is too slight.

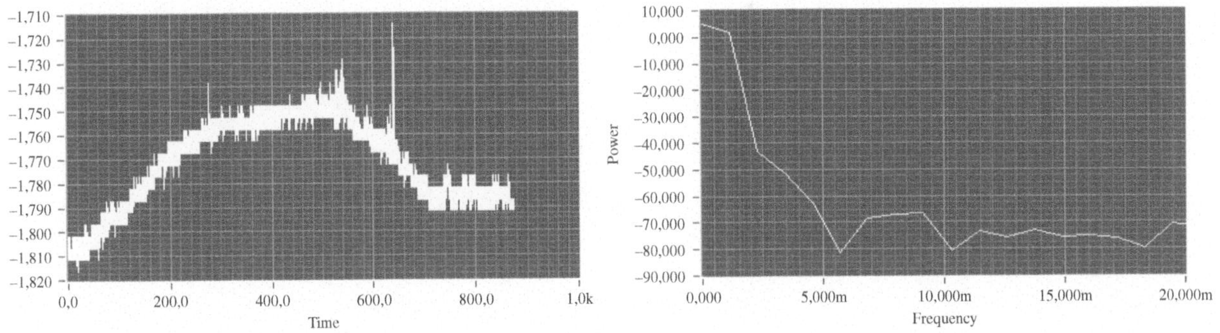

Figure 5.18 CW radar measurements away from leak: 0,1V p-p signal (sec – V) and spectrum (Hz – dB).

REFERENCES

Bimpas M. (2004) Development of Specialized Microwave Sensors- Continuous Wave Radar with Phase Coherence, phD thesis, Athens, October 2004.

Bimpas M. and Uzunoglu N. (2005) Development of a Broadband E-Shaped Ground Penetrating Microstrip Radiator Combining Microstrip Resonators and a Slotted Ground Plane, accepted to be published in the Microwave and Optical Technology LettersBalanis C.: 1989, Advanced Engineering Electro-magnetics, John Wiley & Sons, Inc., pp. 149–151.

Chapter 6

Measurements with the WATERPIPE GPIR system and the LEAKING system

Michail Petsios, Matthaios Bimpas, Michail Sofras, Nikolaos Uzunoglu and Angelos Amditis

The present chapter is the official report providing the description for the results from the prototype Ground Penetrating Imaging Radar (GPIR) that was developed in the framework of the "WATERPIPE" Research Project.

It actually includes some indicative results emanating from both versions of the implemented prototype systems, as well as some information for the leaking radar system that has been integrated with the WATERPIPE system. It gives a good overview of the measurements carried out in the framework of the WATERPIPE project

INITIAL MEASUREMENTS USING THE WATERPIPE SYSTEM

The methodology followed for the evaluation of the GPIR was the following:

(1) In order to assess the performance of the system, we carried out measurements in the lab and at a test site in Athens, after having checked out the performance of our algorithms in simulations. The antennas were measured inside an anechoic chamber, the Printed Circuit Boards (PCBs) were tested using lab equipment, while the algorithms were intensively investigated through intensive simulations using simulated data.

(2) After finishing the first set of lab measurements, the development team of ICCS proceeded with the first version of the integrated prototype system.

(3) The system was tested in two different sites.
   - The first one is the test site made in the NTUA campus,
   - Additionally, we carried out some measurements in EYDAP, which is the water utility of Athens city.

(4) After assessing the measurements there, we did perform an extra set of testing for the algorithms, by making use of raw data emanating from another commercial GPR system.

(5) Afterwards, we ended up with the enhanced version of the GPIR prototype system. Field testing of this prototype can be separated in two stages, measurements without any objects underground and measurements with pipes at different depths. During the testing many changes were made, not only to the hardware of the system (filters were added to avoid interference) but to the software as well (spline approximation in order to improve resolution accuracy).

©2010 IWA Publishing. *Integrated High Resolution imaging Radar and Decision Support System for the rehabilitation of WATER PIPElines.* Edited by Matthaios Bimpas, Angelos Amditis and Nikolaos Uzunoglu. ISBN: 978184333719. Published by IWA Publishing, London, UK.

Figure 6.1 Picture of the system at the measurement site.

Preliminary lab measurements

Some initial measurements were carried out in laboratory environment, by using only dielectric or metallic pipes/objects, in order to assess the performance of the produced GPIR.

The measurement setup is given in the following Figure 6.2, while some indicative results are presented in Figure 6.3.

As it is obvious from the Figure 6.3, there is an imaging capability from the proposed monitoring device; the red coloured sections in the graphs actually represent the existence of a differentiation caused by the metallic material. The next step from the research team was to point out the capability of the system to operate efficiently in the underground media (not the air as shown in the figures above) and for non-metallic objects.

Figure 6.2 Measurement setup for air and metallic/dielectric objects.

Measurements with the system in the test site (NTUA/EYDAP)

Ability to perform 3-D imaging in the underground media

The reconstructed quantities are the spatial distribution of the electric field and/or the polarization current. The plots which are presented in this paragraph show the reconstructed electrical field in 3-D below the antenna receiver block. The reconstructed images are derived from measurements of plastic water pipes buried in soil at a depth of approximately 1 meter up to 1.2 meters (max).

- Resolution in the x and y directions is 10 cm.
  - Images which are shown cover a x-y area of 1×1.2 meters.
- Resolution in the z direction is 7.5 cm.
  - Images show 40 samples which corresponds to a depth of 3 meters.
- Depth direction is upwards (transducer surface is below)

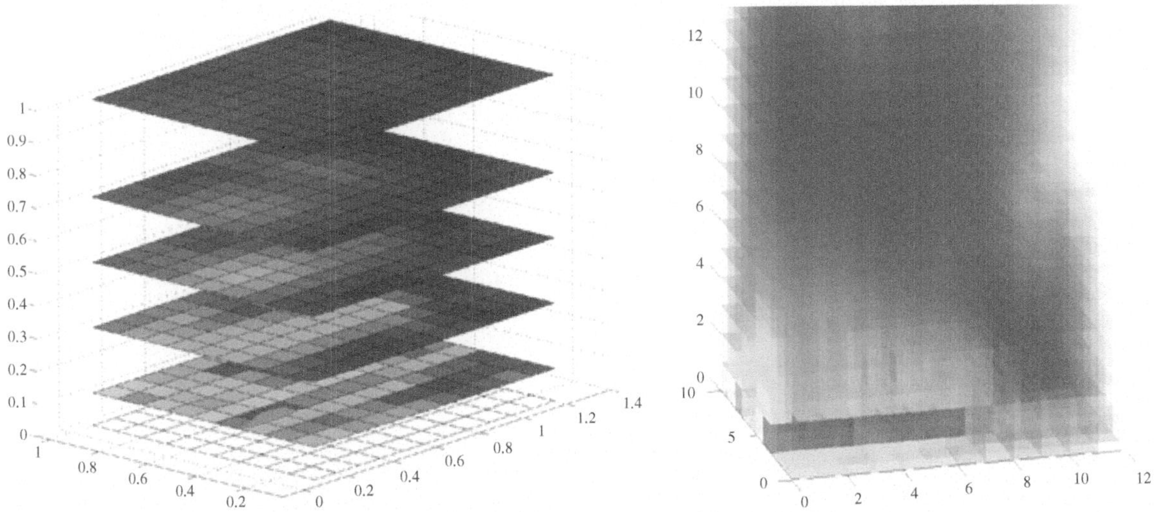

Figure 6.3 Reconstructed field for the copper tube (as shown in the left photo of Figure 6.2) and the cylindrical metallic pipeline (right photo of the Figure 6.2).

In the following Figure 6.4, we can see a first large echo at the surface, which should be our transmitter signals directly received by our receiver array. At a depth of 10 samples, which equals to 75 cm, we can see the pipeline. Our receiver block is at the edge of the pipe. In our effort to analyse the obtained image, we conclude that there is obviously a wrong recognition of the depth and a rather bad quality of the obtained 3-D images, maybe due to a malfunction of the hardware components of the GPIR. These are the main reasons why we considered essential to go for a refinement of the system.

At the following figure, we can see some cuts in the x-z plane, where the existence of an object is quite clear. Depending on the cut, the depth varies from 0,75 cm up to 1 meter that is quite close to real pipeline. It should be noted that the provided results are given in terms of the total number of samples (not depth), due to the fact that the obtained images were much more clear and comprehensible when presented like this.

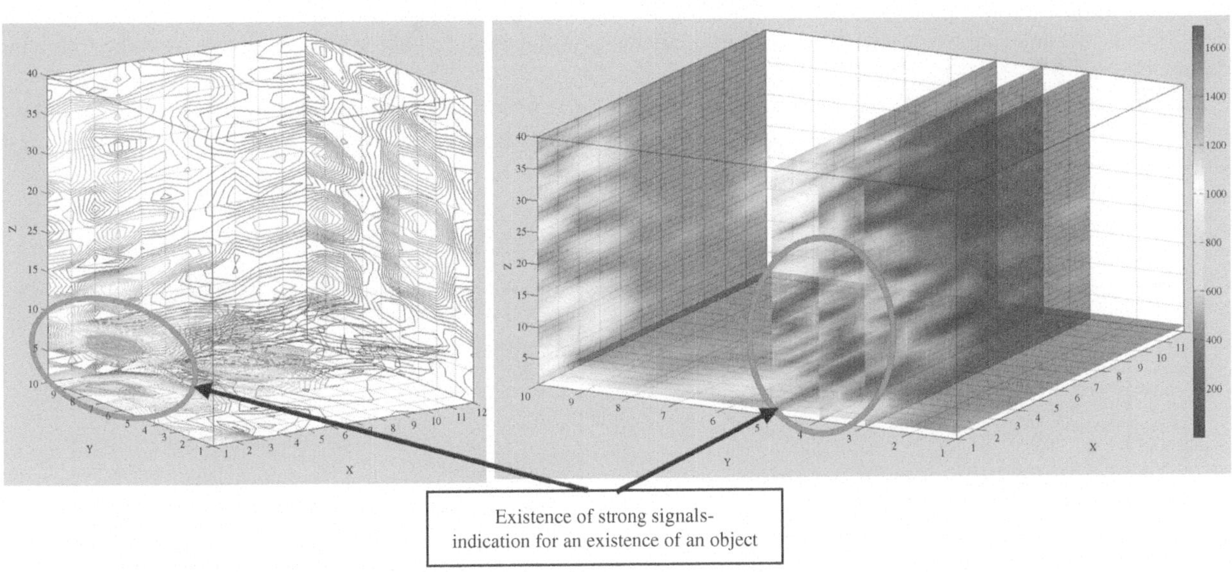

Figure 6.4 (Left) 3-D image of the pipeline buried in the test site of NTUA (Right) Cuts in the x-z plane of the 3-D image of the pipeline.

The next measurement also took place in EYDAP at which we can discriminate an object (please see Figure 6.5), which in this case is the plastic pipeline, detected at the depth of 1.1 meter. Please keep in mind that in this case we made use of depth as the z-axis, due to the fact that the imaging algorithm was capable to distinguish and perform an imaging of the object. This is mainly due to the fact that the scattered signal was rather strong and clear; the underground material may be the reason for this (dry soil that has smaller attenuation compared to the moistured soil). However the accuracy/resolution of the system can not be considered satisfactory, since the geometry of the pipeline is not reflected at all. On the other hand, the depth of the pipeline was correctly identified (approx. 1,1 m), as well as the width of the pipe (approx. 0.3 m nstead of the actual size of 0.45 m).

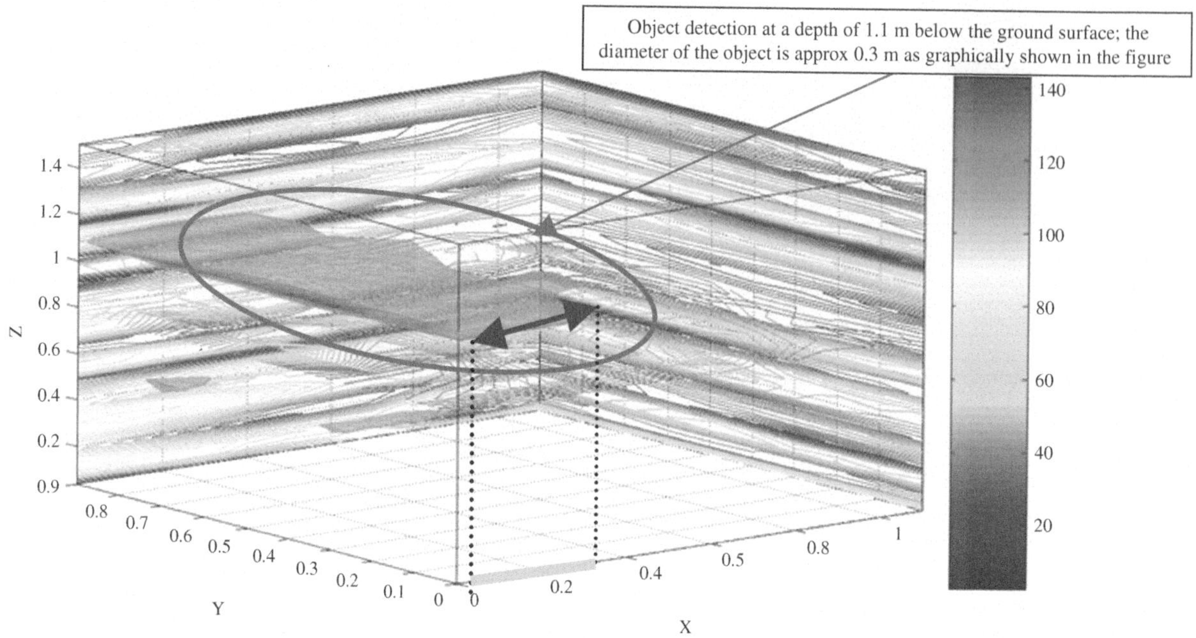

Figure 6.5 3-D image of a plastic pipeline at the depth of 1.1 m in EYDAP test site.

Testing of the performed algorithms

The imaging algorithm of ICCS has been attempted to be tested by making use of data from PipeHawk's radar systems, but it was difficult to do that, due to a totally different data geometry. Thus, we made use of another commercial radar system raw data and we proved the remarkable efficiency of the produced algorithms as shown from the following Figures 6.6 and 6.7.

Data are acquired using a 500 MHz antenna for a drain pipe beneath a concrete floor. The pipe is buried about 0.75 meters beneath the surface, and the "squiggly" pattern near the surface is caused by the wire reinforcing mesh within the concrete slab.

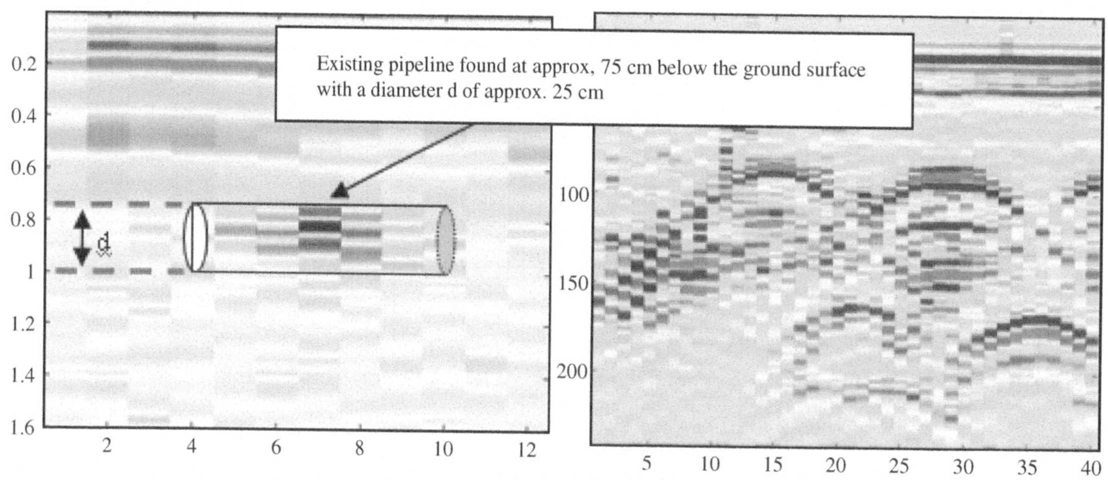

Figure 6.6 (Left) 2-D data representation of a pipeline buried in 75 cm with a diameter of 40 cm. Vertical axis is in meters (Right) GPR Data from Geophysical survey systems, Inc. Vertical axis is in cm.

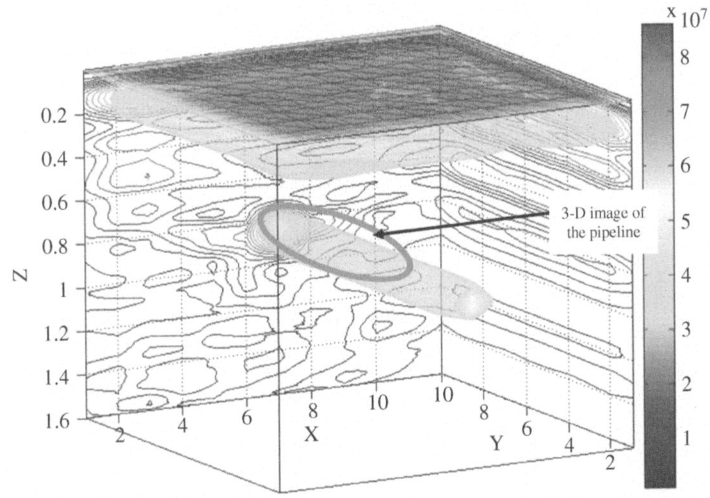

Figure 6.7 2-D data representation of a pipeline buried in 75 cm with a diameter of 40 cm.

The sampling frequency f_s of the acquired data was 8 GHz. A photo taken from the measurements carried out with the enhanced version of the WATERPIPE radar is presented in the Figure 6.8.

Figure 6.8 Photo of the field tests in Athens.

The following figure shows the application of the 3D reconstruction algorithm on these data. Since the developed reconstruction algorithm is inherently three dimensional, data were replicated along the y-direction in order to generate a 3D data set. These images show the excellent ability of the algorithm to depict

By using these equations, we obtain the 3-D graph as presented in the right figure, where we can see a pipeline buried in 0.875 meters, which is quite close to 0.75 meters (actual depth).

As a conclusion, we can say that the testing of the algorithms was successful; however more tests must be carried out before drawing the definite concusions from this innovative equipment. This procedure led the development team to proceed with a refinement concerning the hardware parts of the system, as it will be shown in the next chapter of this report.

REAL MEASUREMENTS BY USING THE ENHANCED VERSION OF THE WATERPIPE GPIR

Field testing can be separated in two stages: measurements without any objects below ground and measurements with pipes at different depths. During the testing many changes were made, not only to the hardware of the system (filters were added to avoid interference) but to the software as well (spline approximation in order to improve resolution accuracy).

As a first result, the following figures present the output of the system using Matlab, firstly without any objects below ground and in the second plot with a metallic pipe buried at 1.5 meters.

It is obvious that in the first plot (please see Figure 6.9), the amplitude of the electric field as reconstructed by the algorithm seems to be uniformly distributed as expected since there is no object below ground. However, in the second plot, the user can easily conclude that there is a discontinuity at 1.5 meters.

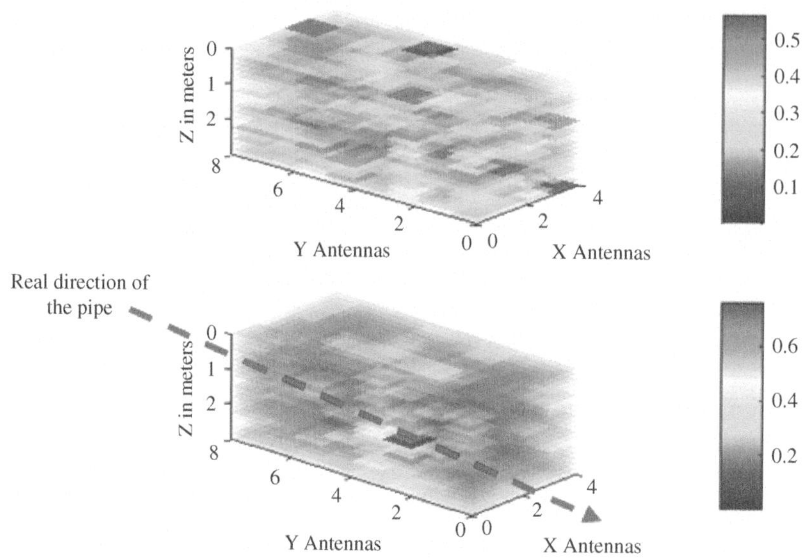

Figure 6.9 System output with no pipe and with pipe at 1.5 meters depth.

As another measurement, a pipe at 0.5 meters depth was placed in order to check the algorithm's ability to estimate the depth. As reported in the next Figures 6.10 and 6.11, the system seems to realize the different depth efficiently.

The final test was with the pipe at 1 meter depth with the system turned 90 degrees. At this measurement, the result is not so clear but an experienced user can conclude the existence of a pipe, the depth and the pipe's potential direction. This performance deterioration was expected since at this direction the system's accuracy is deteriorated due to the number of antennas (4 at x-direction and 8 at y-direction).

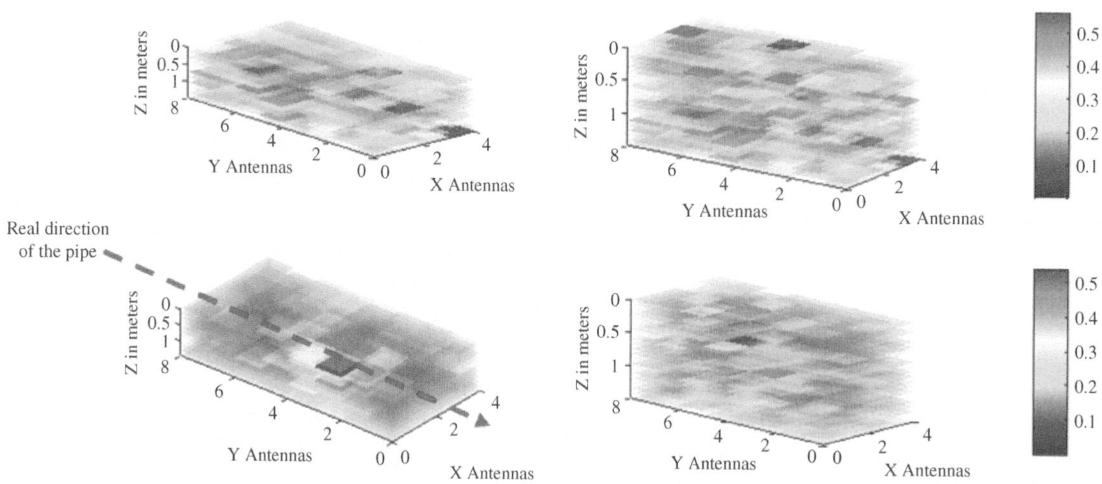

Figure 6.10 (Left) System output with no pipe (top) and with pipe at 0.5 meters depth (bottom) and (Right) the system is turned 90 degrees.

Finally, in order to decrease the resolution cell (increase accuracy), splines approximation was applied to the output of the radar. The output of this processing seems like having 15×7 antennas instead of 8×4.

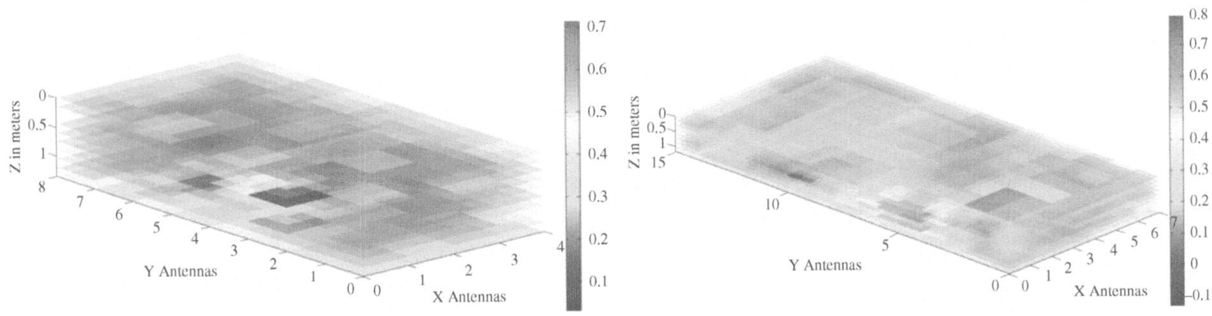

Figure 6.11 System output with pipe at 0.5 meters depth without spline approximation (left) and with spline approximation (right).

MEASUREMENTS BY USING THE LEAKING SYSTEM AT A REAL WATER LEAKING CONDITIONS

Test site measurements

Description of the test site

After many measurements in various test sites, it was considered essential to test the CW radar also in real water leaking conditions. The following Figure 6.12 provides a rough sketch that shows the layout of the pipelines and the measured points. The actual leakage point is also marked on the map.

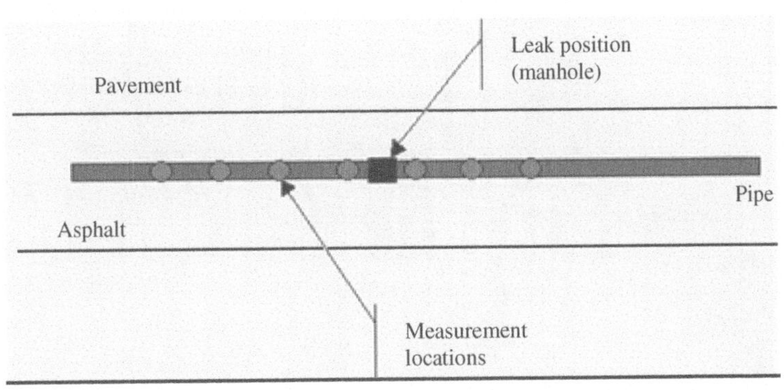

Figure 6.12 Measurement map, in which the measurement points are marked.

Measurement results

The returned signal was recorded for a period of one up to three minutes. The next figures show the results of the sampled data, as a time-domain signal. The horizontal axes of the graphs represent signal amplitude in mV, while the vertical axes represent time. Figures 6.13 and 6.14 depict the case of no leakage at the measurements 3.7 meters and 1.8 meters before the exact leakage point respectively. The criterion that leads the user to this conclusion is not only the maximum fluctuation of the received signal voltage, which does not exceed the value of 45 mV and 20 mV (peak to peak) respectively, but also the signal waveforms, which are not periodical at all.

On the other hand, the results are much better as the radar system approaches the water leakage point, since the maximum signal amplitude fluctuation value exceeds approximately 200 mV (peak to peak), as shown in the following Figures 6.15 and 6.16. At the same time, the received signal turns to be rather periodical and sinusoidal (without any noise). In Figure 6.15 (measurement 0.2 meter before the exact leakage point) the presented part of the received time-domain signal is an ascending part of a sinusoid, while in Figure 6.16. (measurement 0.25 meter after the exact leakage point) the received signal is a descending section of a sinusoid. Thus in both cases, the criterions are fulfilled leading the system operator to the straightforward conclusion of the water leaking existence. It is obvious that the sampling time of 200 sec is not enough to see the total signal period that is approximately 1200 sec. This means that the obtained Doppler frequency shift is less than 1 MHz, due to the very small motion caused by the water leak.

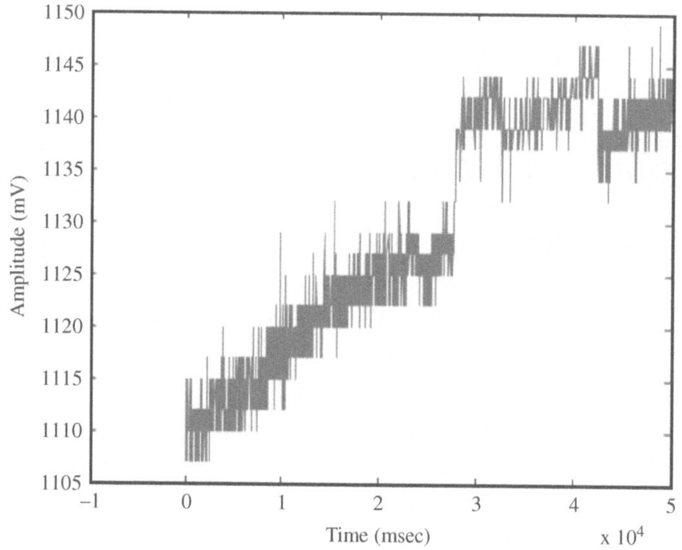

Figure 6.13 Signal 3.7 m before the leak. Characteristics: 45 mV peak-to-peak.

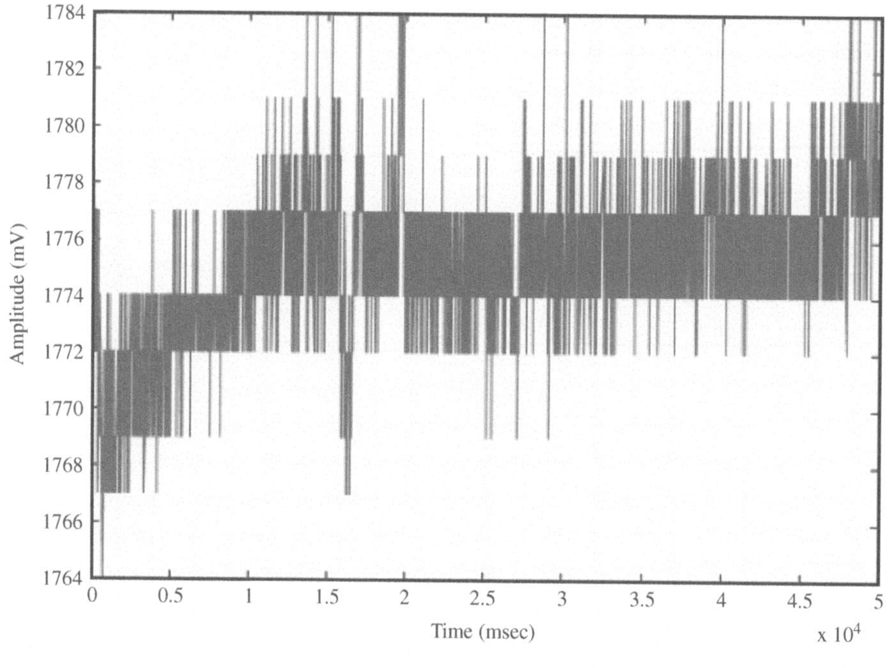

Figure 6.14 Signal 1.8 meters before the leak. Characteristics: 20 mV peak-to-peak.

In the following Figure 6.17, the measurement 0.7 meter away from the actual leakage point is presented. The signal is a little noisy, but still remains periodical, while the signal amplitude is reduced at 75 mV peak to peak. There is a water leakage indication, hence the system user should make further measurements around this area, in order to find out whether this is the main leakage point or it is close enough to this point (as applies to this specific case). It is very important to emphasize that in cases of leakages this is the most usual case that the system user met.

Finally at Figures 6.18 and 6.19 some indicative measurements away from the leakage point are presented.

In the first case (1.15 meters away), the amplitude of the signal (approximately 280 mV) may lead the user to draw the wrong conclusion of water leaking existence. However, the signal waveform characterized by intense signal displacements, is a typical case of operational malfunction. After taking an extra measurement over the same point, the screenshot represented the typical non water leaking case.

In the second case (2.7 meters away from the leak), all the detection criterions lead the system user to the conclusion of non water leakage existence.

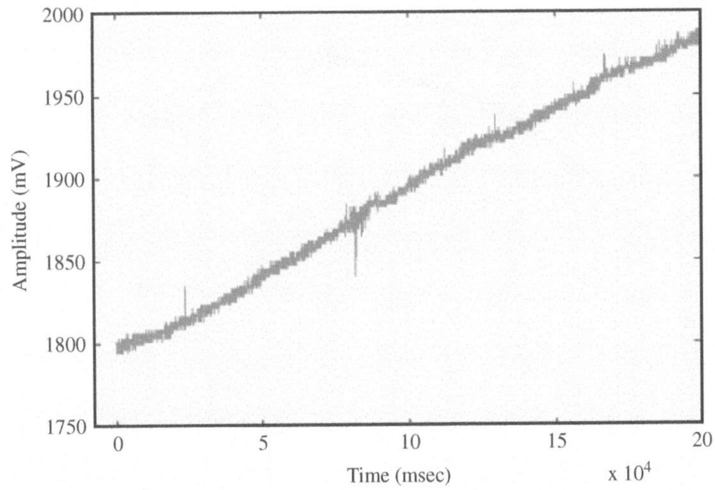

Figure 6.15 Signal 0.2 meter before the leak. Characteristics: 200 mV peak-to-peak.

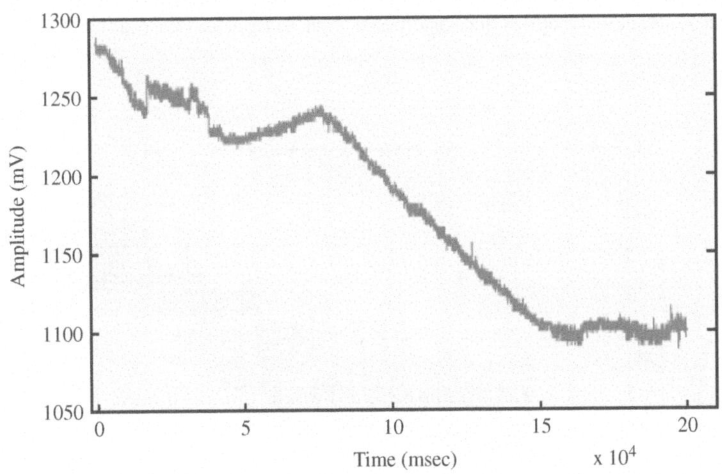

Figure 6.16 Signal 0.25 meter after the leak. Characteristics: 210 mV peak-to-peak.

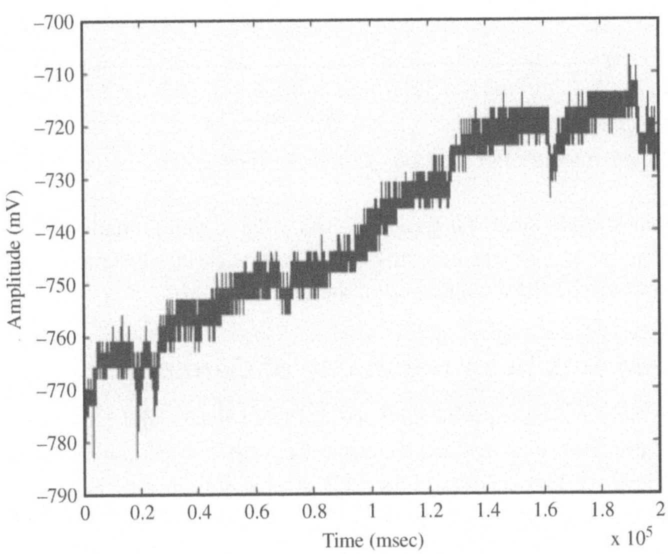

Figure 6.17 Signal at 0.7 m after leak. Characteristics: 75 mV peak-to-peak. The cause of the Doppler shift has faded significantly.

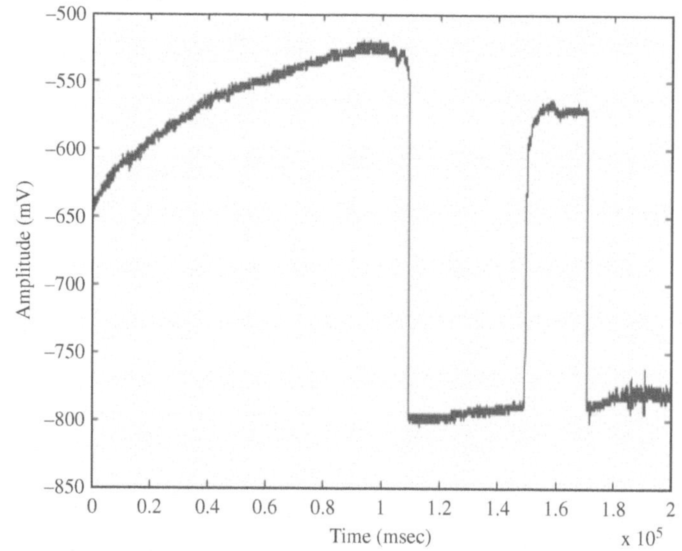

Figure 6.18 Measurement at 1.15 meters after leak. Characteristics: 280 mV peak-to-peak with signal displacements due to operational malfunction.

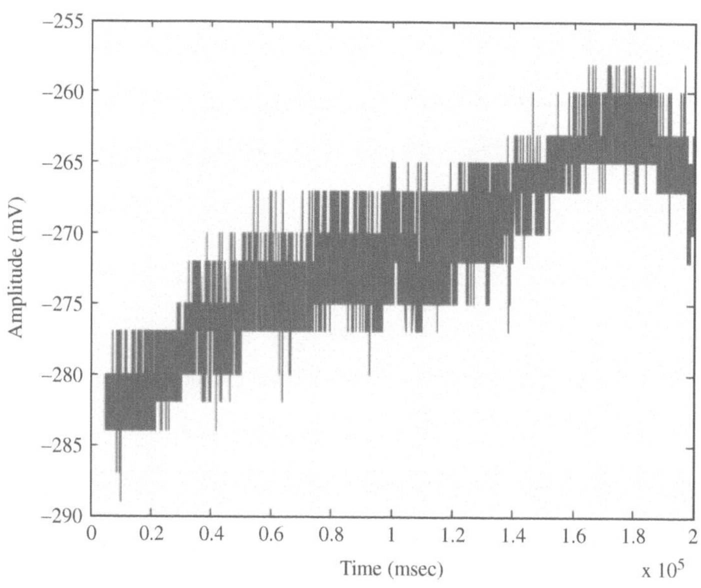

Figure 6.19 Measurement 2.7 meters after the leak. Characteristics: 32 mV peak-to-peak, none detected motion.

There was no measurement straight over the manhole, since the cover is made of metal. Therefore, it cannot be penetrated by microwave radiation. More measurements were taken at further distances, but all of them showed a weak, noisy signal, that cannot be confused with leak detection signal.

Comparative measurements in an urban city of Germany

After having tested the CW radar system and verified its ability to detect and locate accurately the water leakages, excessive comparative measurements in Chemnitz (Germany) were carried out between the proposed system and prevalent commercially methods. In this section, some indicative results are presented in demanding cases of water leaks. A multi-point correlator was compared to the CW radar system. SEWERIN, which is a company that offers products and services for water supply utilities, was supervising the whole measurement procedure. It should be mentioned that the correlator was chosen by SEWERIN systems operators, since it was considered to perform better than all the other instruments for water leaking detection.

In Table 6.1, the main characteristics of fifteen different leakages in real conditions, along with the results of the measurements using the correlator and the CW radar system, are given. In the second and the third column of this table, the pipeline material and the type of soil (underground propagating media) are given respectively. The measured distance (in meters) from a reference point resulted from the tested equipment (correlator and CW radar system) are presented in the fourth and fifth column of Table 6.1. The measurements given with the bold characters indicate the results, which were considered satisfactory from the SEWERIN system operators. The dash where appear in some of the cells of these two columns means that the specific equipment did not show any indication for water leakage. Finally the sixth column shows the actual distance from the reference point in which the leakage was found when the excavation took place by SEWERIN, after the end of the measurement procedure.

Table 6.1 Comparative measurements between correlator and the proposed CW radar system.

No	Material	Soil	Correlator (m)	CW (m)	Actual (m)
1	steel	clay, wet	13.79	–	14
2	steel	clay, wet	–	0	0
3	steel	sand, wet	21.25	20	21.4
4	steel	clay, wet	31.67	32	32.2
5	cast iron	stone, wet	3.34	1.8	3.4
6	cast iron	clay	9.61	10	No leak
7	cast iron	clay, wet	10.96	10.8	10.9
8	cast iron	clay, wet	18.42	17.4	17.4
9	cast iron	clay	27.72	23.4	27.4
10	cast iron	sand, dry	3.56	3.8	3.6
11	PVC	clay, wet	2.79	2.9	2.9
12	PVC	stone, wet	2.02	2.40	2.0
13	PVC	clay, dry	3.62	3.7	3,6
14	PVC	sand, wet	–	17	17.1
15	PVC	sand, dry	0	–	2.0

In all the above cases, the water pipelines were buried in the ground at about 1.1 meters up to 1.6 meters and the leakage size varied from 1 m^3/h up to 4 m^3/h. Judging from these results, both the equipment had an excellent performance in ten (10) cases out of the total fifteen (15) measurements. This percentage may be considered very satisfactory, especially for experimental equipment such is the developed CW radar system. Especially in cases of PVC pipelines, the CW radar had four out of five satisfactory indications, while the correlator system performed satisfactory only in three out of five cases.

CONCLUSIONS

The System uses a fixed antenna array and is designed to be light weight in order to conduct measurements easily. The field measurement results are encouraging compared to other systems having 3-D imaging capability; however extensive tests are needed to be carried out in order to assess the accuracy and resolution of the presented radar system.

In any case, based on the measurements presented at this report, it is obvious that yet we miss any information regarding the existence of cracks or water-leaks. Last but not least, the system performs with some uncertainty in the detection and the location of the pipelines.

As far as the results coming from the CW radar are concerned, these are satisfactory for detecting Doppler shift existing due to water leak motion. Compared to the other mentioned water leaking techniques (correlators, acoustic devices, etc.), CW radar exhibits superior performance, since it is immune to environmental conditions (weather, street noise, etc.) and its efficiency does not depend on the pipeline material (especially PVC), as shown in the previous paragraphs.

On the other hand, the presented CW radar system cannot perform efficiently, in case where pipelines are buried deep in the ground or in cases of very damp soils, which introduce very high attenuation, due to the high operating frequency.

REFERENCES

Kechribaris, C.N., Nikita, K.S. and Uzunoglu, N.K. (2000) "Reconstruction of Two-Dimensional Permittivity Distribution using an Improved Rytov Approximation and Non-linear Optimization, Journal of Electromagnetic Waves and Applications, MIT.

Lymperopoulos, D.G., Bimpas, M., Stratakos Y., Karanasiou, D.I. and Uzunoglu, N. (2003) 'Detection of Water Leaks in Supply Pipes by Using Microwave Ground Penetrating and Thermography Techniques', 3rd International Conference on Non-Destructive Testing of the Hellenic Society for NDT, Vol. 1, pp. 330, Crete, 15–17 October.

Sorensen, J.D. (1993) 'Reliability-Based Inspection Planning for Structural Systems', Reliability and Optimisation of Structural Systems, V, Proceedings of the IFIP WG7.5 Fifth Working Conference on Reliability and ptimisation of Structural Systems, Takamatsu, Kagawa, Japan, 24–26 March, P. Thoft-Christensen and H. Ishikava, eds. North-Holland, Amsterdam, pp. 31–46.

Zaridze, R., Bit-Babik, G. Karkashadze, D., Jobova, R., Economou, D. and Uzunoglou, N.K. (1998) "The Method of Auxiliary Sources (MAS) Monograph", Tbilisi-Athens, National Technical University in Athens.

Zaridze, R. (1998) "The Method of Auxiliary Sources and Scattered Field Singularities(Coush'cs)", *Journal of Electromagnetic Waves and Applications*, **12**, 1491–1507.

Chapter 7

Design, implementation, testing and refinement of the expert system, the data base, the data manager and the geographic information system GIS module

Stefanos Camarinopoulos, Mata Frondistou, Guenter Becker, Vasilis Kalidromitis and Corrado Sana

INTRODUCTION

The purpose of this chapter is to set out the guidelines design, testing and refinement of the Expert System (ES), Database (DB), Data Manager (DM), and GIS. As the design has advanced with the clear longer-term goal of module integration of the Decision Support System (DSS), both it and the implementation methodology reflect an explicit preparation for the integration stage of the software development life-cycle. Following detailed software design phase, this task is based on the testing and refinement guidelines dictated by the European Space Agency (ESA) Software Engineering Standards.

For the development of the DSS Modules on Leakage, Structural Reliability and Concentration of Contaminants, the above tasks are related to:

- Design of the overall model architecture, description of the the data-flow control, design and production of system interfaces and coding of the detailed design into the software prototype;
- Testing of the modules, integration of modules, functional testing and refinement of coding based on the analysis of results.

The following paragraphs report on the results and achievements of the above activities.

For the three sub-modules (Structural Calculation, Leakage and Concentration of Contaminants) the following structure has been adopted in order to provide both determinist and stochastic calculations and relevant outputs:

According to the above structure the deterministic modules can run separately and provide results. This is particularly important for testing purposes with an important decrease of calculation timing.

SYSTEM ARCHITECTURE

The design of the system architecture is bound to be guided by the end goal: the successful integration of all modules into a working system, as shown in the Figure 7.1 below. Based on a rough description of the methodology, one starts with the requirements and specification of all modules and decides how their functionality is to be realised, on a

 Integrated High Resolution imaging Radar and Decision Support System for the rehabilitation of WATER PIPElines. Edited by Matthaios Bimpas, Angelos Amditis and Nikolaos Uzunoglu. ISBN: 978184333719. Published by IWA Publishing, London, UK.

module-by-module basis. Next comes the design of the interfaces between the modules: this step ensures that a common communication protocol is agreed upon between modules, such that independent and in-parallel work can be carried out between development teams (at the same time facilitating the substitution of modules by functional equivalents).

Rather than segmenting the architectural description of the system in hierarchical levels, the description of architecture is best performed considering two view-points

- The logical viewpoint is a synopsis of the abstract specification of all modules and their interactions, as first shown in the Requirements and Specification deliverable
- The physical/technical viewpoint sets out the physical details of the system architecture including, among others, development and target platforms and the development tools to be used.

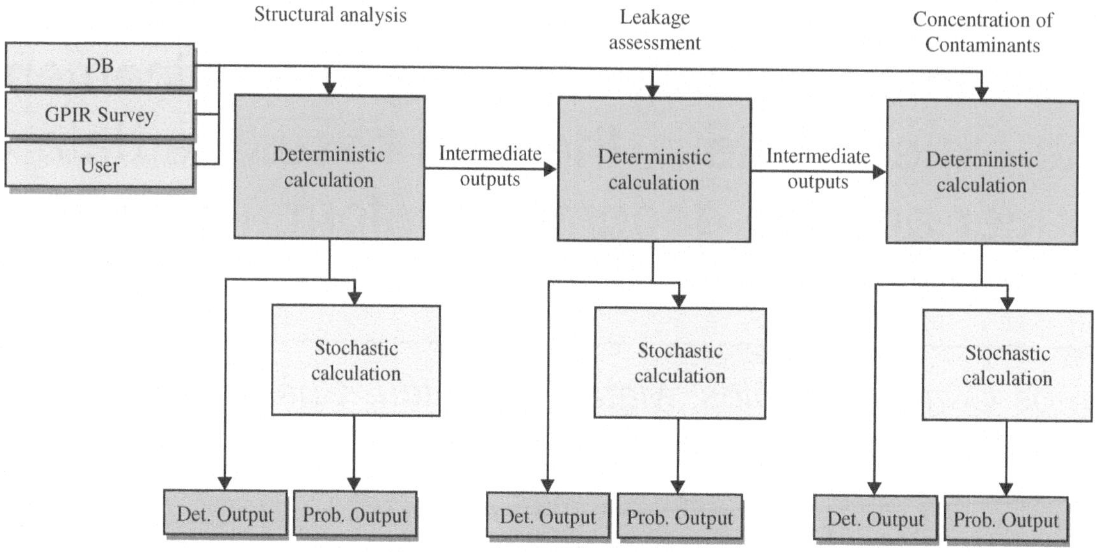

Figure 7.1 The Structural Reliability module architecture.

Logical system view

The flow of information required for the processing of a particular user query has been described in the Requirements and Specification deliverable. The following figure describes the interactions between the modules.

Figure 7.2 outlines the global system architecture. User requests to the specific system may be of the form:

- When will a specific pipeline segment, of given diameter, wall thickness, material, joint type, age, soil type, depth of cover, etc. inspected for leakage 3 years ago, reach a point that the probability will be more than 50% that more than 30% of the water conveyed will be lost to leakage?
- In the absence of any leakage control in the above segment at what point in time will there be a probability of more than 50% of structural failure?
- If a decision is being reached to rehabilitate in 2010 a specific failed pipeline segment what will be the net present value of costs of the best rehabilitation option?
- What are the non-economic consequences of pipeline failure and repair, rehabilitation and re-inspection cost?

Such requests should be addressed to the system either in textual or a more friendly graphical form.

The ES module should act as an intelligent intermediary between the user and the system, thus breaking up a user query into appropriate subparts, feeding them into the corresponding software modules (the reliability assessment, rehabilitation lifecycle cost assessment and prioritization module) while synchronising their communication, summing up the individual replies and finally constructing a comprehensive answer for the user. Thus, the ES acts more of a coordinating system rather than a classic diagnostic ES.

Although the exact internal architecture of the ES was finalized since the design phase, it grossly comprised the following components (see Figure 7.3):

- a user interface UI, responsible for the input of user queries, their corresponding output and the provision of a basic navigation to the data base.
- an inference engine which decides on actions to be taken regarding the execution of a user query, combining domain and system knowledge from the Knowledge Base (KB).
- a query manager, responsible for the decomposition of queries to subparts, appropriate handling of each part and synthesis of the final answer.

- a DM, mainly responsible for the provision of an implementation-independent data interface. Among the specialised data manager tasks, are the enforcement of specific data integrity and access control mechanisms, default value management and scenario-based data handling.

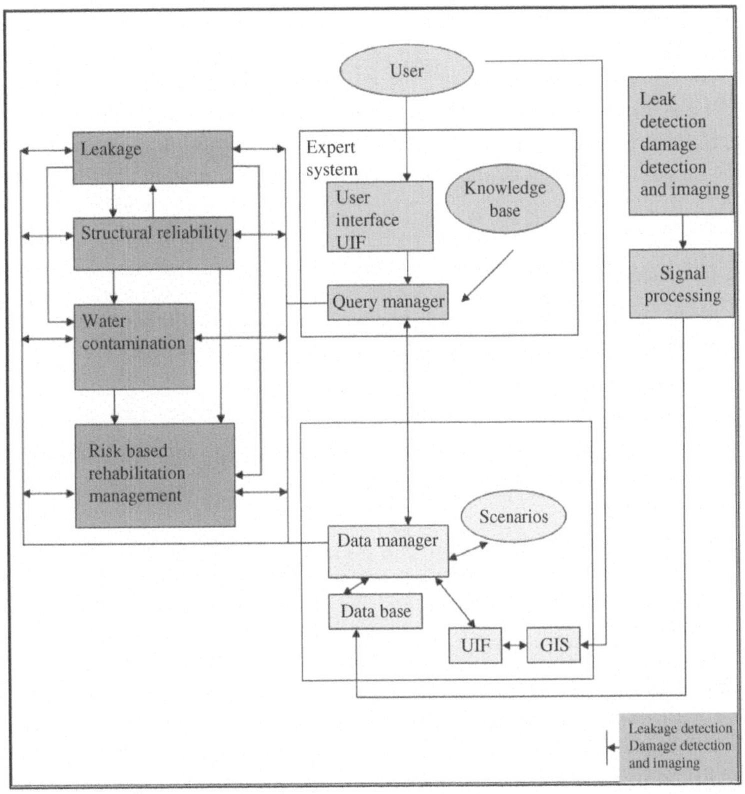

Figure 7.2 Integrated system for the rehabilitation management of the trunk and distribution water mains.

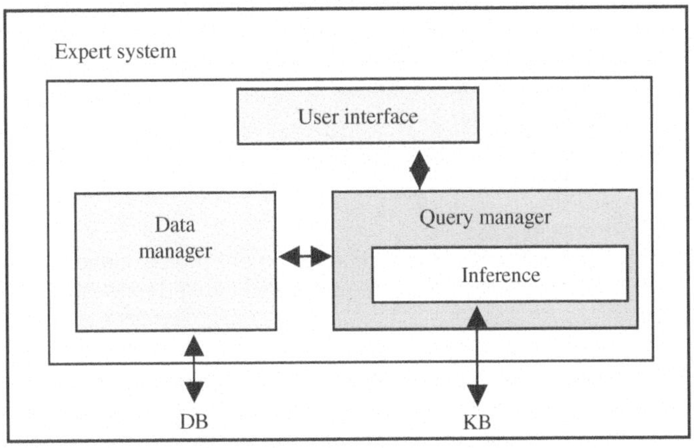

Figure 7.3 Expert System description.

The KB contains domain and system (rules concerning interaction of software modules) knowledge, while the database constitutes the integrated data repository servicing all the different modules with appropriate data for their processing.

As a subpart of the DSS, the ES along with the DB and KBs are data-intensive in nature: their primary function is to store, process and provide data to the other modules as well as provide integration and coordination of user interaction with the system as a whole. Thus, a formal or semi-formal (as often analysis & specification practice suggests) description of this subpart's functioning mainly entails a corresponding specification of the information content the subpart handles (i.e. a conceptual database schema).

The physical system view

The logical system view does not describe how the modules have been implemented or how communication between these modules are realised. These details are described below.

Programming languages

The structural reliability, Leakage, Contamination and the Risk Based Rehabilitation Management modules have been developed in JAVA and Visual Basic by RISA and TECNIC. Activation of the structural reliability and Risk Based Rehabilitation Management are triggered by the Expert System. These modules access the database for reading and writing data via the Data Manager. The methods included in the module Application Programming Interfaces (APIs) are developed in java.

Module interaction/database integration

The structural reliability can have access to the database via calls to the Data Manager with a common mechanism, i.e., the Data Access Object (DAO) and ActiveX Data Object (ADO) Microsoft technology.

Graphical User Interface (GUI) design

Human Computer Interface (HCI) designing is the process of designing interface software so that computer systems are efficient, pleasant, and easy to use and do what people want. The following points make it easy for the user to operate the system.

The design of the UI was linked to the study of HCI in connection to the results from the interview analysis.

The UI is implemented in Java and provides a graphical environment with which the end-user can retrieve current and historical data from the database. It is foreseen that the interface provides warnings in case of abnormal situations and allows the end-user to examine different scenarios for hypothetical situations. The ES is also implemented in Java and it is responsible for the co-ordination of the system modules.

The UI is interconnected with a GIS package (MapObjects) to offer additional functionalities, i.e. for the visualisation of the water network and to allow an easier way to select particular regions for the determination of their structural reliability and the graphical display of results.

Platform issues

The common development environment used in the WATERPIPE project consists of the following parts:

Operating system:	Windows XP service pack 2
GIS:	MapObjects for Java version 2.0
Development tools:	Visual Studio version .NET (Visual C++ and Visual Basic)
	Jbuilder2008
	AGXIS
Workstation:	Pentium-class PC
Database:	Microsoft Access in the development phase, since it is included in Microsoft Office without additional licence fees and is used extensively.

In principle, for the local development of single modules, different platforms may be used. These platforms must, however, be compatible with the common development environment to ensure the seamless integration of local implementations into the final system.

DATA DESCRIPTION

This section offers a concise description of the information content of the ES and the data and KBs. It must be noted that at this level no clear distinction is drawn on "knowledge" and "data". Rather, information units crucial for the operation of the DSS, irrespective of their specific nature, are identified and recorded in a formal yet practical and comprehensible way.

On conceptual modeling

Computers can understand file records, e.g. small portions of the magnetic surface of a hard disk containing values in binary form, whereas humans can handle, reason and interact in natural language on entities like bank accounts, hotel rooms, airflight seats, etc. Between the two extremes, intermediate representations are possible, each at a particular abstraction level which filters out unnecessary details of lower levels and allows a confined description of the part of the real world that is modeled. In a DB context these representations are called schemata (singular: schema) and the tools available to construct them are called data models. It is a well established practice in the design of a database to follow 3 distinct phases, corresponding to the following schemata:

(1) *physical schema*: it is a precise description on *how* the data are actually stored and accessed (file organisation, collections of index structures and corresponding algorithms).
(2) *logical schema*: it is the next higher level of abstraction, describing *what* (instead of how) data are stored in the database and the interrelations that exist among them.
(3) *conceptual schema*: it is the abstract user view of the part of the real world at hand, capturing its semantics. It is mainly a collection of entities, their properties and their interrelationships, i.e. the static aspects of the modeled application.

This fragmented design approach ensures the so-called data independence: the ability to modify the schema at some level, without necessarily altering the schemata at the higher levels (e.g. without the need to rewrite the application programs).

Moreover, a conceptual schema is usually rich of high-level constructs which constitute a good communication medium between the user and the application designer. Often performed in refinement cycles, this design phase constitutes a formal specification of the information content of the application, at a level abstract enough to be user comprehensible but also rigorous enough to form the basis of the logical schema definition. Such an approach minimizes potential errors in the analysis process and ensures proper specification of the data requirements, portability, serviceability and application viability, since any design modifications at this level (rather than code patching) are likely to be seamlessly incorporated through the intermediate steps to the final software product.

MODULE SPECIFICATION – INTEGRATION

From a process-oriented perspective, this Section describes the functionality of the Coordinating system as a flow of actions initiated by the user and subsequently passed through the appropriate software modules, with the expert system having the overall coordination, as it is shown in the Figure 7.4.

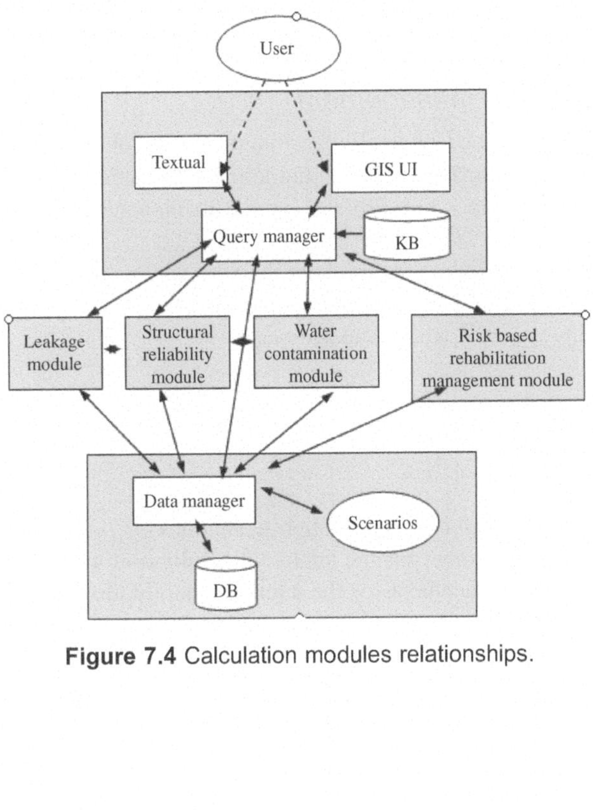

Figure 7.4 Calculation modules relationships.

1. The structural reliability module

According to the developed methodology the three sub-modules (Structural Calculation, Leakage and Concentration of Contaminants) have been developed adopting the following structure in order to provide both deterministic and stochastic calculations and relevant outputs:

According to the above structure the deterministic modules can run separately and provide results. This is particularly important for testing purposes with an important decrease of calculation timing. The developed UIs include three main types:

(1) **UIF for data input**. They show the values from DB or default values and also allow the modification of concerned data.

(2) **UIF concerning the management of intermediate data and/or assumptions affecting calculations** (for example combination of loads to apply to the selected pipes. These kinds of interfaces are mainly used for testing and/or for standalone running of the modules for deterministic calculations.

(3) **UIF including output data.** They include tables and graphics reporting the main results.

The below reported schema, as reflected by Figure 7.5, shows the UIF structure:

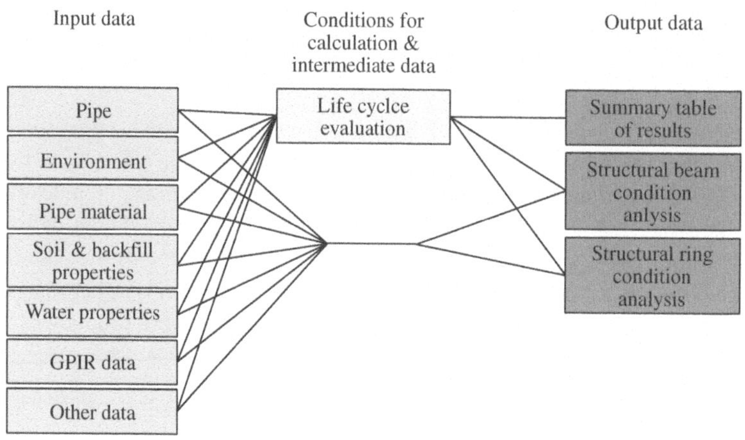

Figure 7.5 The Structural Reliability UI structure.

Since the calculation modules give results whose reliability depends on consistency of input data, the developed UIFs fully support step by step calibration procedures, giving the opportunity to define different testing scenarios (defining input data and load combinations) and check not only final but also intermediate results.

2. The Risk based rehabilitation management

The purpose of this module is to identify the technically feasible rehabilitation methods and to estimate the Net Present Value (NPV) of the different lifecycle costs of these rehabilitation and replacement options by taking into consideration not only the initial investment costs of pipes but also the annual maintenance costs for the life of the pipes and the salvage value of the pipe as well.

The Rehabilitation Options and Cost Module obtain from the user and/or the data base basic information about the parameters necessary for the functionality of the procedure such as:

- number of cross section investigated where leakage occur
- position
- maximum probability of failure
- existing pipe diameter
- depth of cover
- working pressure
- the expected service life for the applied rehabilitation techniques
- cost per meter with service replacement for the applied rehabilitation techniques
- cost per meter with no service replacement for the applied rehabilitation techniques

Some additional input data are provided directly from the user or the expert system such as:

- the time horizon
- the discount rate
- the inflation rate
- the service replacements

The module returns a list of all the technically feasible rehabilitation and replacement options. Then risk is calculated by combining the output of failure prediction models with the output of cost assessment models.

The rehabilitation options have been identified and include replacement (renewal) or renovation (relining) that can be structural or non-structural. Moreover, rectification of the structural damage can be local, not involving the whole pipe (structural repair).

The impact of the above options has been assessed on life cycle direct and social costs and the time required before the next intervention has been determined.

The developed risk-assisted approach consists of evaluating the different rehabilitation options using risk analysis. It involves calculation of the risk (in monetary terms involving the Net Present Value (NPV) of aggregated costs) for the different rehabilitation scenarios. The scenario with the lowest NPV of costs is retained.

3. The expert system (ES)

The Expert System is responsible for the co-ordination of the system modules and the interaction between the end user and the system.

The user interface – UI

The DSS is intended for use mainly by non-computer-literate technicians of water utility companies. Therefore, emphasis was laid on four aspects regarding the design of the UI:

- *ease of use* – all users should be able to reach effectiveness and ease of use without any particular training.
- *guidance* – the user should be able to find his way in the interface with minimum effort, achieving easy and correct use of commands, thus increasing reliability and effectiveness of use.
- *error anticipation and prevention – All users (even experienced ones) inevitably make mistakes. A good system should be able to guide the users for* the easy and correct use of its commands so as to avoid errors. However, in the case of error the system should be able to deal with it effectively, issue the appropriate error-message and avoid disruption of the operation.
- *integration* – the DSS consists of separate components which communicate via external routine call mechanisms. It is important for efficiency and effectiveness of use that the user should be able to work in a conceptually unique and coherent environment, handling transparently inter-module communication via appropriate mechanisms and operating system calls. Moreover, interaction with the user should be avoided if execution of operations could be achieved without any user input.

Table 7.1 summarizes the main principles of human-computer interaction. These principles together with simplicity were the main guidelines for the design of the UI.

Table 7.1 Basic principles of human-computer interaction.

1.	***Consistency*** – in presentation of tasks, presentation of information, layout design.
2.	***Compatibility*** – user's expectations should match with reality of the interface. The design should be based on user's experience.
3.	***Adaptability*** – the interface should adapt to the user in several ways (speed of work, skill level, etc.). The user (rather than the computer) should be in control.
4.	***Economy*** – the user should achieve the desired operations with the minimum number of steps.
5.	***Guidance*** – the interface should guide the user through the tasks to be completed and not control him.
6.	***Structure*** – The design should reduce complexity by classifying similar components together (according to the user's mental model of breaking down the particular problem).

A typical user interaction with the WATERPIPE DSS involves simple data entry/browsing or posing of a query and visualization of the results. Achieving this, in line with the afore-mentioned design principles, leads to the specification of a standard menu-driven interface, assisted with the utilization of a GIS component for the alternative specification of a water pipe subset upon which the user query must run. An indicative query processing example is outlined below:

- **Step 1.** The user issues the query: Assess the structural reliability of a specific pipeline segment, of given diameter, wall thickness, material, joint type, age, soil type, depth of cover, etc. inspected for leakage 3 years ago, that more than 30% of the water conveyed will be lost due to leakage.

Suppose that the user wants to assess the structural reliability of a pipe segment of a pipe (see Figure 7.6) with pipe segment length = 6 m, start node id = 10 end node id = 20 and leakage results by pipe segment:

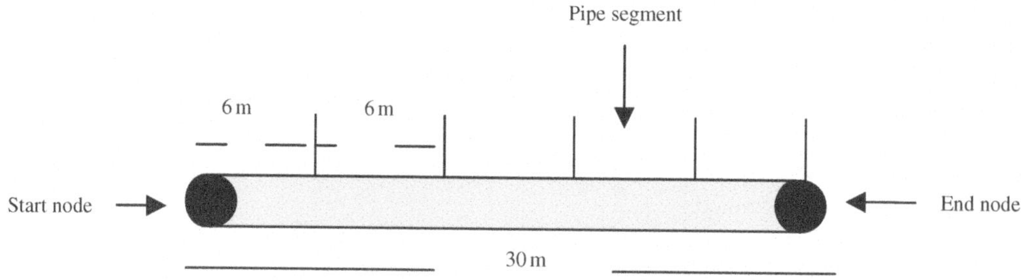

Figure 7.6 Water pipe deviated in segments.

The user has to indicate the pipe segment which can be done with the use of a textual menu or alternatively, the user is faced with a map of the pipe-network through which a pipe is graphically selected. Suppose that the user selects textual a pipe segment as shown in Table 7.2 (e.g. pipe segment with id = 5)

- **Step 2.** The query manager internally transforms the query, invokes the data manager to supply corresponding data and calls appropriately the reliability assessment module. This module runs several times for each pipe segment separately.
- **Step 3.** The query manager reads the results from the database and constructs a query answer.
- **Step 4.** The result to the query (a sequence of time steps, starting at the year the pipe segment was laid, each of which is associated with a probability of failure) is fed as a chart into the UI (see Figure 7.7).

Table 7.2 Leakage level per pipe Segment.

Pipe id	Pipe segment id	Leakage survey date	Start node	Distance (m)	Ave Leakage level
1	1	14/05/2003	10	6	0 (no leakage)
1	2	14/05/2003	10	12	0 (no leakage)
1	3	14/05/2003	10	18	1 (low leakage)
1	4	14/05/2003	10	24	3 (High leakage)
1	5	14/05/2003	10	30	2 (Medium leakage)

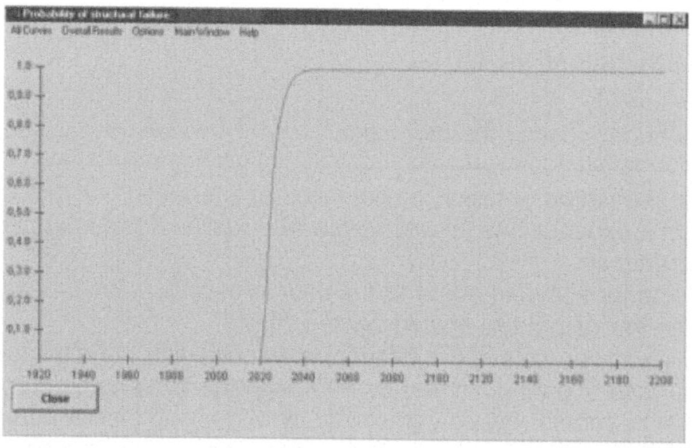

Figure 7.7 Typical result to a reliability assessment query.

According the reliability results a decision can be reached to rehabilitate a specific failed pipeline, thus the net present value of costs of the best rehabilitation option is calculated.

Step1. The use selects the pipe segment

Step2. The query manager internally transforms the query, invokes the data manager to supply corresponding data and calls appropriately the rehabilitation lifecycle cost assessment module .

Step3. The query manager reads the results from the database and constructs a query answer.

Finally getting input from reliability module all failed pipeline segments priorities for rehabilitation are assigned based on the non-economic consequences of failure and repair, rehabilitation and re-inspection cost. In Table 7.3, the failure probability of segment 5 is given for various time points.

Table 7.3 Failure probability of segment 5.

Pipe id	Pipe segment id	Time point	Failure probability (%)
1	5	2020	0,20
1	5	2025	0,40
1	5	2030	0,70
1	5	2035	0,80
1	5	2040	0,90
1	5	2045	1
1	5	2050	1
1	5	2055	1
1	5	2060	1

GIS interface specification

As already stressed, the purpose of the GIS component is to assist the user in the process of pipe selection and in general provide a friendly linkage between a graphical representation of the pipe network and the database. Therefore, it should typically provide the following features:

- a graphical selection tool (either by simple mouse click, within rectangle selection or within arbitrary polygon selection
- zoom in and zoom out tools
- a pan tool
- an "identification" tool (presentation of database attributes for the graphically selected nodes)
- a data entry tool via simple graphical selection
- multilayer presentation (the ability to graphically combine different map layers and corresponding annotation in order to facilitate user navigation)

The query manager

The role of the query manager is three-fold:

- interpret user queries
- decompose them in different parts and address them to the appropriate modules
- compose a result returned to the user

To accomplish these tasks, the query manager (i) makes use of its embedded inference engine and the KB in order to schedule proper sub-query processing actions (ii) interacts with the data manager so as to acquire appropriate data for query processing (iii) using the API (see section 3.5), translates the user request to appropriate subroutine calls.

The data manager – DM

The DM provides suitable data to caller modules, according to the module's particular view. It is itself envisioned as a module consisting of subroutines that access the database. Further responsibilities of the data manager are the following:

- the support of "what-if" scenarios. A what-if scenario is essentially a user query with the additional requirement of storing the input parameters and the results. Therefore, to implement such a scenario, a copy of the database instance that represents all its input parameters is required along with the corresponding results and any other information (comments, date, etc.) supplied by the user.
- the ability to assess similarity among stored queries. Identifying similarities (including a measurable way to define them) between the input of two distinct queries could save a significant processing overhead. In other words, having processed different reliability scenarios for different parts of the network, one could infer that the reliability behavior of a specific pipe in a particular network part would be similar to that of an already processed part, as long as the two input conditions are similar. Defining similarity is not a trivial engineering task. It may well be possible that "similarity" is reduced to strict equality of values as the behavior of the models is not linear to their input. So, scenario similarity is something that was further explored in the implementation phase.
- handling of default values. Populating the database by entering values explicitly for each pipe is certainly a tedious a time-consuming task. Two scenarios are considered to overcome this:
    - assigning of values according to an augmented look-up table, containing several frequently encountered pipe types. Therefore, the user may select a certain type and automatically populate database information like diameter, working pressure, pipe density, material, modulus of elasticity, etc.)

    - a similar approach is the use of a look-up table containing pipe types according to their geographic distribution. This approach is appropriate for determining geographically varied parameters, such as frost load, traffic load, etc. A combination of the two approaches should suffice for the population of most default pipe parameters in this sense.
- Handling of null values. In case the database contains null values for parameters essential for the computations of the reliability or rehabilitation modules, the data manager issues appropriate alerts and can optionally assign values for these parameters according to predefined rules.

The API concept

The final system consists of the packaging of individual software modules and the availability of the decision support process to the end user through a neatly defined and well-organised interface. The clearest benefit of good integration is that self-contained black-box modules can be plugged in and out with minimal overall system modifications – this is especially important for sound maintenance.

To transfer this goal into a functional system, every software module must inform all outside callers of how it may be called, what it expects as input and what it produces as output. This is achieved through the Application Programming Interface (API), which is a set of subroutine specifications that the software module makes public to other software modules within the system.

The software modules can only be accessed through their API and, as long as a caller software module uses the API, its programmer need not worry whether the called module has been implemented in a particular way.

From another viewpoint, a software module can be thought of as a library, which has numerous internal functions but only allows the programmer to use some of these functions. In that case, the specification of the library functions that the programmer may use serves as the library's API.

The cornerstone of a successful integration effort is to ensure that the APIs are neat, well defined and adhered to.

There are a few central rules to the effective use of the API concept.

- It is built before writing code.
- The API concept is actually structured in two layers, the software module API and the subroutine API
- The subroutine API is the description of how a specific subroutine may be called (in the spirit of what has been presented above)
- The software module API is the set of subroutine APIs for all (public) subroutines of the module
- The code to be written follows the API specification.
- All changes are documented and all other software modules are accordingly informed (however, changes can be kept to the absolutely minimum if the initial understanding of module interaction is sound).

Chapter 8

Field testing report for the Decision-Support-System (DSS)

Nicola Bazzuro, Kalman Koncz, Sophie Kanellopoulou, Stefanos Camarinopoulos, Stuart Hamilton and Bambos Charalambous

TESTING OF THE DSS IN AMGA, ITALY

Software application

The Project Administrator (Figure 8.1 – left) allows the user an easy project management using the commands Open, Create, Delete, Change and Copy Project. By clicking on Open Project the user can open the GIS Explorer (Figure 8.1 – right), which is the core of the user-software graphical interface and the WATERPIPE Explorer.

The GIS Explorer has the typical characteristics of GIS applications. In addition to the classic Pan, Zoom In/Zoom Out, Full Extent, Next/Previous Extent, it presents the following commands:

- Identify: identifies the code of the selected pipe which is highlighted for some time in the network layout. The intersections between two or more pipes are assigned to only one pipe, by clicking on it only this pipe is highlighted but the codes of all the intersecting pipes are displayed.
- Point selection: open the Pipes Editor panel (Figure 8.2) related to the selected pipeline. It is not very simple to select every pipe and sometimes the command does not promptly respond.
- Window selection: Allows highlighting a section of the network, but the command functions are not very clear.
- The Find command allows a search for individual network elements, that is search by address, using geocoder, but it is not easy to use due to the lack of a specific help.

The WATERPIPE Explorer allows the user to display information related to the project elements and presents two menus, File and Edit. The File menu allows data import and export into and from the project. The Edit menu allows users, with reference to the different project data categories, (Networks, Pipes, References, List), to open, copy, delete, assign values, undo or modify an entry related to the recorded data, create new records. Not all the functions can be fully exploited, since the Help cannot be opened and error messages are often in German (this happens also in other software tools).

By opening a record in one of the data categories of sub-categories (Pipes, Backfill Properties, etc., Figure 8.2a) the user can view the related WATERPIPE Editor forms; the pipe forms allow users to view and modify data related to water properties (Figure 8.2b), pipes (Figure 8.2c, Figure 8.2e) and the surrounding environment (Figure 8.2a, Figure 8.2d).

©2010 IWA Publishing. *Integrated High Resolution imaging Radar and Decision Support System for the rehabilitation of WATER PIPElines*. Edited by Matthaios Bimpas, Angelos Amditis and Nikolaos Uzunoglu. ISBN: 978184333719. Published by IWA Publishing, London, UK.

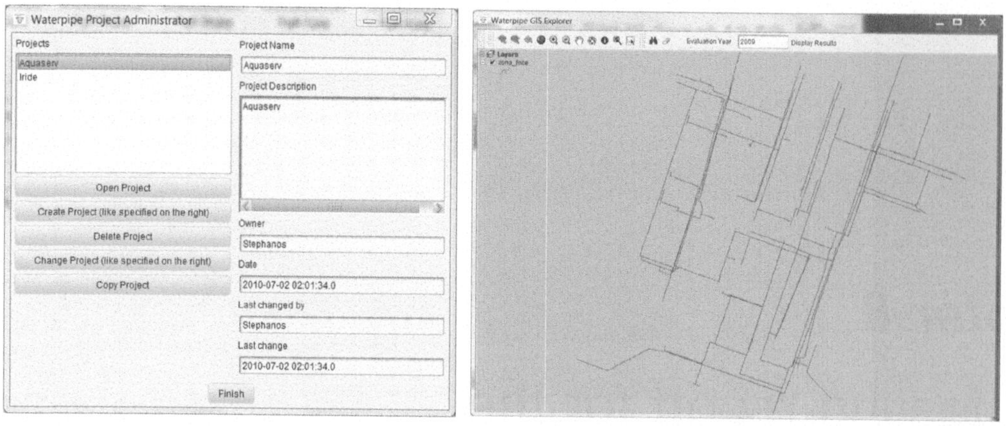

Figure 8.1 (Left) Project administrator main dialog box, (Right) WATERPIPE GIS Explorer.

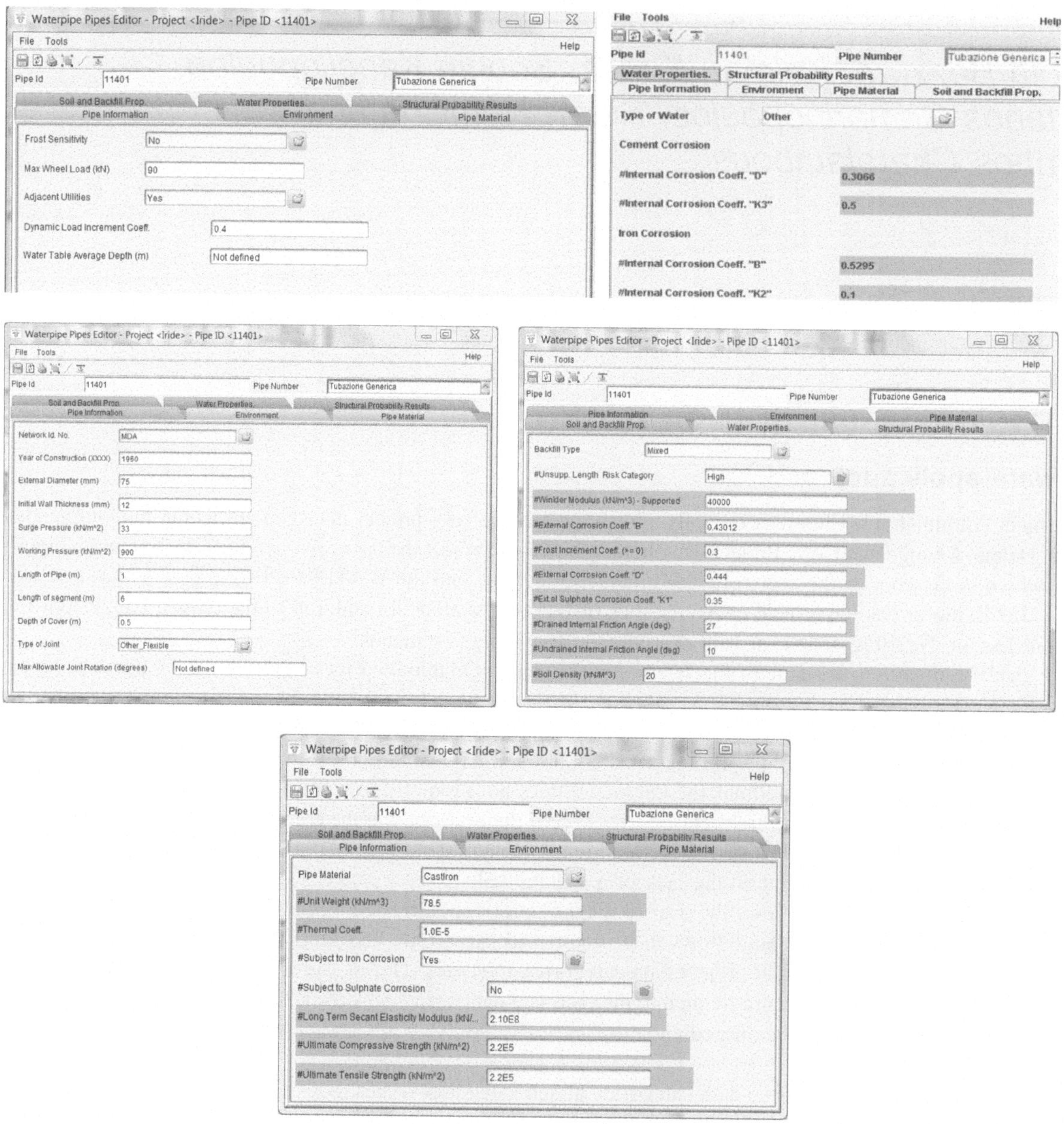

Figure 8.2 Pipes Editor – forms.

From the Pipes Editor toolbar the user can launch the Structural Reliability Module and the Risk Based Rehabilitation Model. The Structural Reliability Module panel (Figure 8.3), that must be opened pipeline by pipeline (and does not display the related ID), includes the input data (modifiable by the user) related to the pipeline (Input Data Selection), the load calculation combinations referred to beam conditions and ring conditions, the time horizon and the two calculation options: Deterministic Calculation and Probabilistic Calculation.

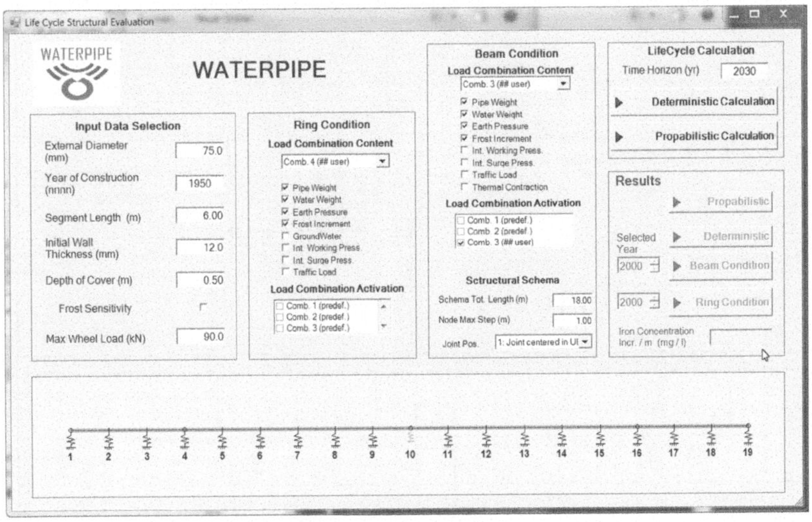

Figure 8.3 Structural Reliability Module.

The option Deterministic Calculation provides the following results:

- Deterministic,
- Beam Condition,
- Ring Condition and
- Iron Concentration

Deterministic (Figure 8.4):

- with reference to Ring Condition, the user can view for every year, from present time to the reference time horizon (the default period is from 2004 to 2029), the number of radial failures and the related position in the network section, the leakage level, the maximum duration (in minutes) of biological contaminations, (for instance from escherichia coli);
- with reference to Beam Condition the user can view the number of horizontal failures, the related longitudinal position with respect to the pipe, the duration of contaminations, the leakage level from horizontal failures and joints. Users can view and save results as .txt files.

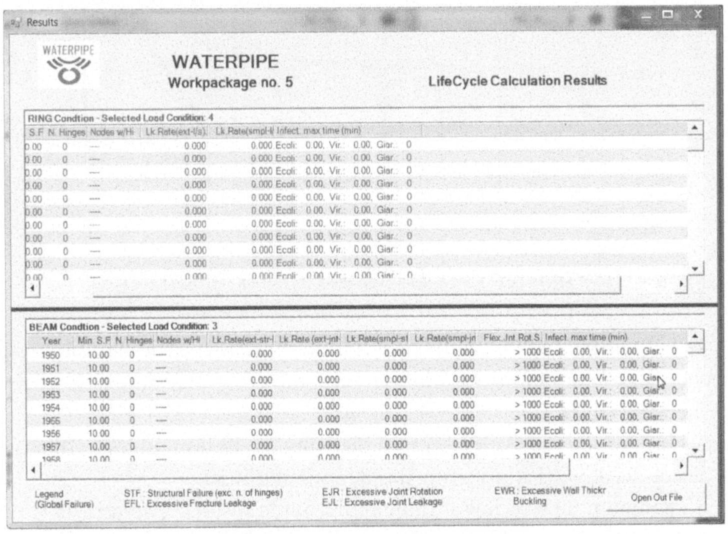

Figure 8.4 Deterministic Calculation – Deterministic results.

Beam Condition (Figure 8.5): the user can view the load diagram and options and, by clicking on Forward, longitudinal bending moments, normal and cutting stresses, strains along the upper and lower generatrixes of the pipe external wall, threshold levels for tensile and compressive stresses and safety factors.

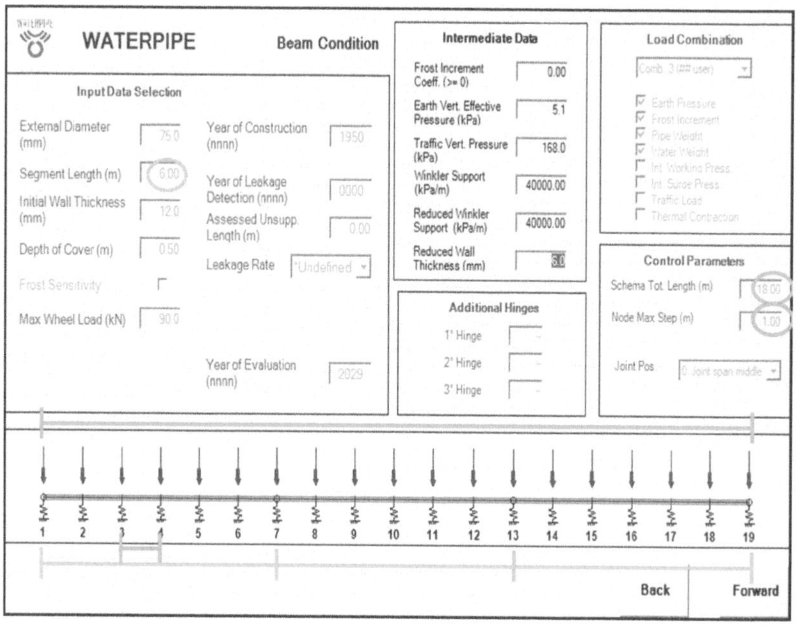

Figure 8.5 Deterministic Calculation – Beam Condition results.

Ring Condition (Figure 8.6): the user can view the load diagram and options and, by clicking on Forward, circumferential bending moments, normal and cutting stresses, strains along the inner and outer pipe wall surfaces, threshold levels for tensile and compressive stresses and safety factors.

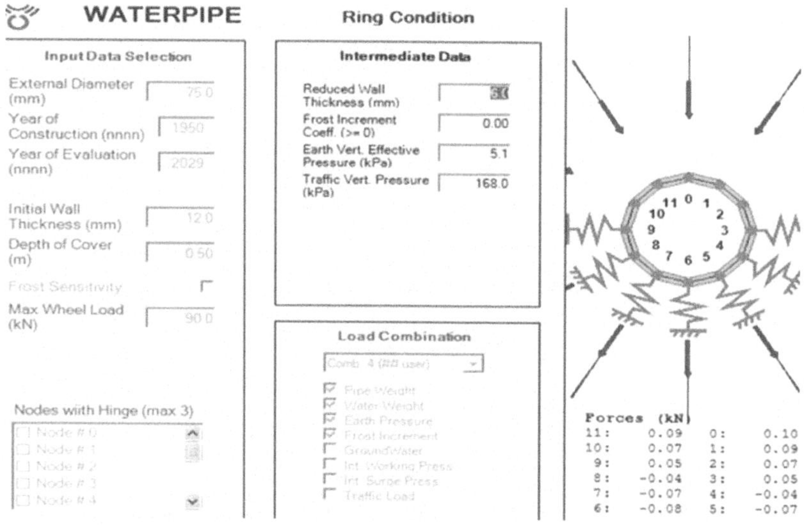

Figure 8.6 Deterministic Calculation – Ring Condition results.

Unfortunately, there isn't a help menu and this can cause misinterpretation of some results, for instance the parameter Min. S.F. (possibly, it indicates the minimum safety factor), the resulting and threshold stresses (two tensile and two compressive stresses), and the safety levels.

Furthermore, it is not very clear which transverse section the Ring Condition refers to (the most stressed, the one with the lowest safety level, or a section defined a priori in the default settings).

The option Probabilistic Calculation takes a few minutes to provide results. The results (Figure 8.7), for the pipeline under survey, different for the Ring Condition and Beam Condition, can be viewed and saved in a .txt format. The results provided for the Ring Condition are the average leakage level, failure and leakage probabilities. The results provided for the Beam Condition are the average leakage levels from horizontal failures (Structural Leakage) and joints, the probabilities of leakage due to failures and from joints and the failure probabilities.

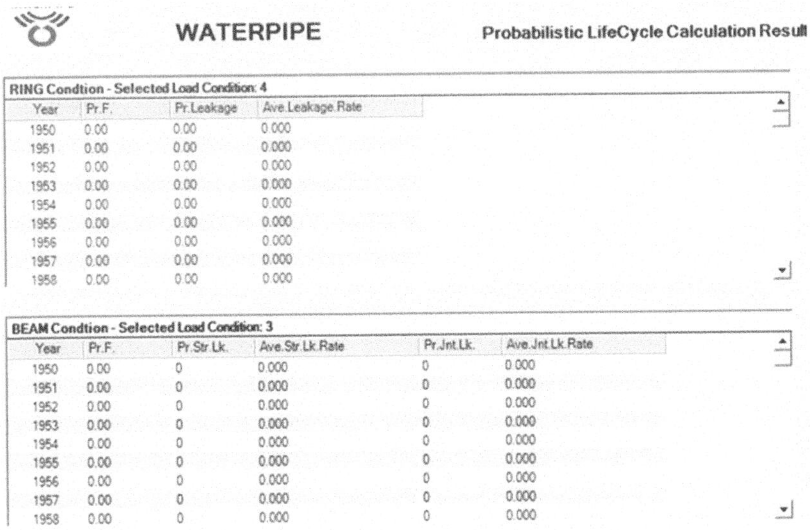

Figure 8.7 Probabilistic Calculation Results.

The Risk Based Rehabilitation Model panel (Figure 8.8),that must also be opened pipeline by pipeline (and does not display the related ID), includes the following input parameters:

- Pipe Data, data related to pipelines (modifiable by users): diameter, material and length (it is not clear why the length default setting is 1 m).
- Damaged Length, percentage of the pipe damaged length.
- Structural Failure Probability, data on the probabilities of failure or ovalization of the tube section (longitudinal cracks), of circumferential failures transverse with respect to the pipe axis, of joint failures. The role of the parameter Evaluation Year is not very clear, maybe it refers to the year in which the failure probability is evaluated.
- Presence of other subsurface utilities in the vicinity of the water network, water leakage, water quality failure due to the presence of biological pollutants or salt irons; problems connected with the pipeline hydraulic capacity.

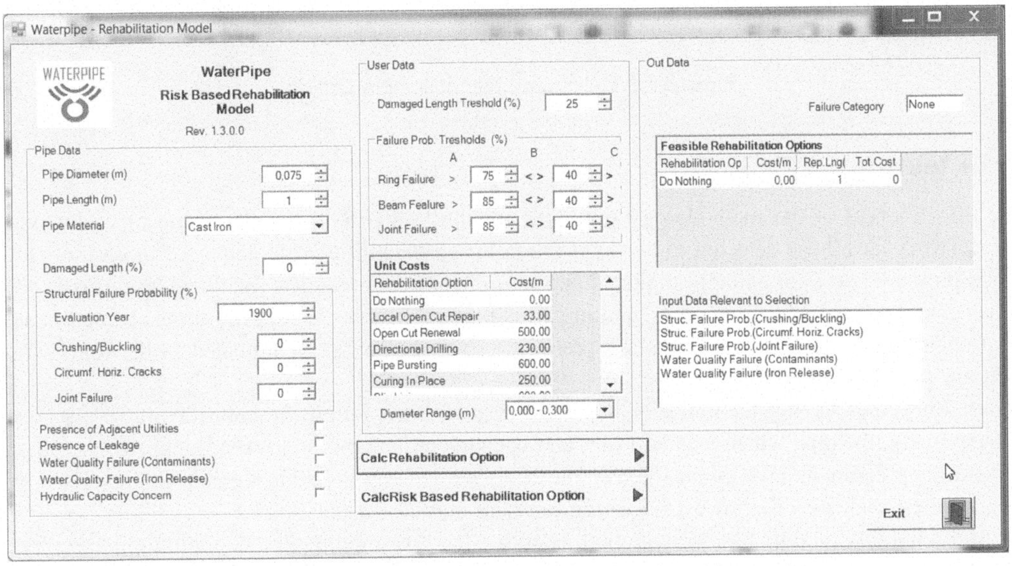

Figure 8.8 Risk Based Rehabilitation Model.

User Data:

- Damaged Length Threshold: according to the instructions provided in the file Wp_Rehab_user_ defaults.txt, is the percentage of the pipe damaged length, above which the pipe is included in the failure category A or B. Anyway, the characteristics of the different failure categories are not very clear.
- Failure Probability Threshold: according to the file Wp_Rehab_user_ defaults.txt, it refers to the two probability values that, for each kind of failure (Crush Buckling, Circular Horizontal Crack, Joint Failure), distinguish the three failure categories, A, B or C.
- Unit costs for each rehabilitation option, Diameter Range function, that is: Do Nothing, Local Open Cut Repair, Open Cut Renewal, Directional Drilling, Pipe Bursting, Curing in Place, Slip Lining, Epoxy Lining, Cement Mortar Lining.

The Calc Rehabilitation Option (11) – please see a screenshot at Figure 8.9 – gives the following results:

- Failure Category refers to the most probable failure that can occur: no failure, Cat. A, Cat. B, Cat. C.
- Feasible Rehabilitation Option:
  - Most convenient rehabilitation option, Local Open Cut Repair, etc.
  - Unit cost of the rehabilitation option.
  - Length of the pipeline section to be repaired.
  - Overall rehabilitation costs.
- List of the input data needed to select the rehabilitation option.

The second calculation option, Calc Risk Based Rehabilitation Option, does not seem to give any kind of results, since there isn't a specific input procedure for the needed data and the output file is not displayed on the screen.

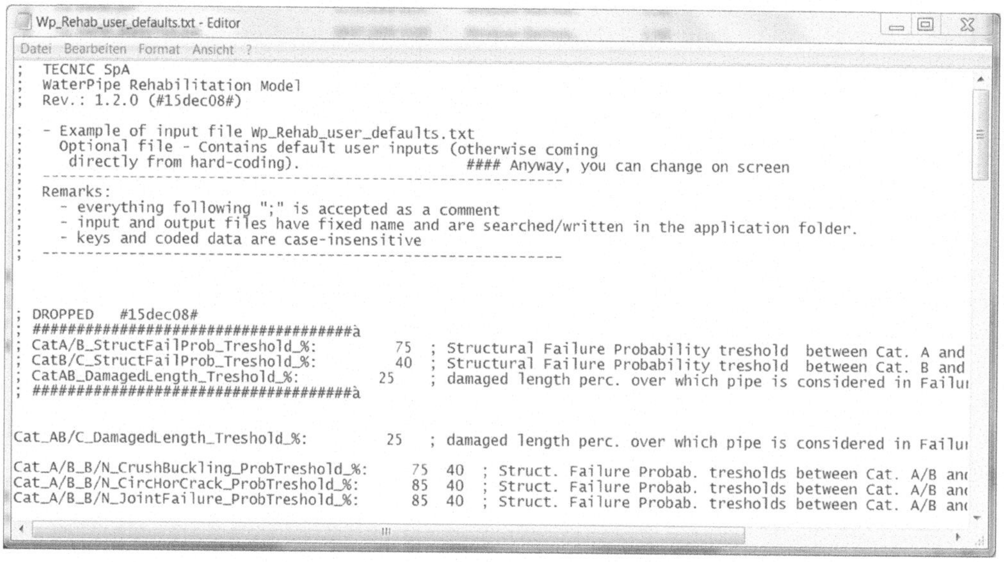

Figure 8.9 Calculated Rehabilitation Option.

Verification tests

To verify the effectiveness of the methodology proposed for the WATERPIPE DSS, some pipes included into Iride Acqua Gas waterwork have been selected to assess if the failure probabilities calculated for those more at risk (due to age, road loads pressure, etc.) are actually higher than the probabilities predicted for those less at risk.

Analysis has concerned only cast iron pipes, and in particular pipe couples having the same characteristics apart from thickness, diameter, year of installation, operating pressure, cover thickness, as indicated in the following tables. The pipes selected are presented in Table 8.1.

The two cast iron pipes having theoretically the highest and lowest failure probabilities have also been analysed, which are, respectively, the pipe which was laid first and is characterised by the lowest thickness, the largest diameter, the highest operating pressure and the thinnest cover, and the pipe which was laid last and is characterised by the highest thickness, the smallest diameter, the lowest operating pressure and the thickest cover.

The Structural reliability module has been applied, adopting the "user-defined" calculation mode set by default. The only non-null results are related to Probabilistic calculation for Ring condition and are analysed in the following sections.

Table 8.1 Pipes selected to evaluate the developed WATERPIPE DSS – Genoa.

Material	ID	Year	Diameter	Thickness	Joint	Cover	Pressure	Seg. Length	Length
CastIron	25616	1950	200	4	Flexible	0.5	600	6	9
CastIron	23328	1950	200	4	Flexible	0.5	600	6	42
CastIron	21179	1950	175	8	Flexible	0.5	600	6	61
CastIron	21692	1950	175	8	Flexible	0.5	600	6	40
CastIron	21686	1950	175	8	Flexible	0.5	600	6	30
CastIron	5067	1950	100	3	Flexible	0.5	900	6	99
CastIron	5210	1950	100	3	Flexible	0.5	900	6	93
CastIron	6763	1950	100	3	Flexible	0.5	900	6	123
CastIron	21594	1950	100	3	Flexible	0.5	600	6	92
CastIron	23234	1950	100	3	Flexible	0.5	600	6	91
CastIron	21670	1950	100	3	Flexible	0.5	600	6	113
CastIron	21597	1950	100	3	Flexible	0.45	600	6	181
CastIron	21224	1950	100	3	Flexible	0.4	600	6	91
CastIron	5212	1935	100	3	Flexible	0.5	900	4	119
CastIron	8085	1935	100	3	Flexible	0.5	900	4	116
CastIron	8076	1935	100	3	Flexible	0.5	900	4	105
CastIron	23328	1950	200	4	Flexible	0.5	600	6	42
CastIron	21692	1950	175	8	Flexible	0.5	600	6	40
CastIron	8076	1935	100	3	Flexible	0.5	900	4	105
CastIron	5067	1950	100	3	Flexible	0.5	900	6	99
CastIron	21594	1950	100	3	Flexible	0.5	600	6	92
CastIron	21224	1950	100	3	Flexible	0.4	600	6	91
CastIron	5087	1935	100	3	Flexible	0.5	900	4	197
CastIron	38931	2009	60	6	Flexible	0.8	600	4	54

Pipes having the same characteristics apart from : Thickness and diameter;
Pipes having the same characteristics apart from : Year of installation;
Pipes having the same characteristics apart from : Pressure;
Pipes having the same characteristics apart from : Cover;
Cast iron pipe "in worst condition";
Cast iron pipe "in best condition".

The following testing scenarios were selected:

- Test Scenario G1: pipe "in worst condition" vs. pipe "in best condition"
- Test Scenario G2: pipe 'having the lowest thickness and the largest diameter' vs. pipe 'having the highest thickness and the smallest diameter'
- Test Scenario G3: pipe 'that was laid first' vs. pipe 'that was laid last'
- Test Scenario G4: pipe 'characterized by the highest operating pressure' vs. pipe 'characterized by the lowest operating pressure'
- Test Scenario G5: pipe 'characterized by the thinnest cover' vs. pipe 'characterized by the thickest cover'

The detailed results of the test scenarios are presented below.

Test Scenario G1 – Pipe "in worst condition" (ID 5087) VS pipe "in best condition" (ID 38931):

RING CONDITION ## **Pipe 5087**

Year	Pr. F.	Pr. Leakage	Ave. Leakage. Rate	Year	Pr. F.	Pr. Leakage	Ave. Leakage. Rate
1935	0.000	0.000	0.000	1954	0.320	0.320	3.672
...	...	...	...	1955	0.378	0.378	3.672
1950	0.015	0.015	5.507	1956	0.378	0.378	0.000
1951	0.033	0.033	5.507	...	...	...	...
1952	0.040	0.040	5.507	2028	0.378	0.378	0.000
1953	0.140	0.140	3.672				

RING CONDITION ## **Pipe 38931**

Year	Pr. F.	Pr. Leakage	Ave. Leakage. Rate
2009	0.000	0.000	0.000
...	...	...	
2028	0.000	0.000	0.000

Result: HYPOTHESIS VERIFIED

Test Scenario G2 – Pipe having the lowest thickness and the largest diameter (ID 23328) VS pipe having the highest thickness and the smallest diameter (ID 21692):

RING CONDITION ## **Pipe 23328**

Year	Pr. F.	Pr. Leakage	Ave. Leakage. Rate	Year	Pr. F.	Pr. Leakage	Ave. Leakage. Rate
1950	0.000	0.000	0.000	1984	0.690	0.285	7.447
...	...	...	...	1985	0.753	0.303	6.107
1974	0.030	0.000	0.000	1986	0.823	0.373	7.524
1975	0.030	0.000	0.000	1987	0.825	0.373	0.000
1976	0.040	0.010	9.160	1988	0.948	0.425	6.979
1977	0.053	0.023	9.160	1989	0.960	0.438	9.160
1978	0.053	0.023	0.000	1990	0.975	0.438	0.000
1979	0.118	0.053	6.870	1991	0.995	0.458	9.160
1980	0.180	0.080	6.107	1992	0.995	0.458	0.000
1981	0.283	0.098	9.160	1993	0.995	0.458	0.000
1982	0.335	0.120	8.482	1994	1.000	0.458	0.000
1983	0.438	0.183	7.084				

RING CONDITION ## **Pipe 21692**

Year	Pr. F.	Pr. Leakage	Ave. Leakage. Rate
1950	0.000	0.000	0.000
...	...	...	
2028	0.000	0.000	0.000

Result: HYPOTHESIS VERIFIED

Test Scenario G3 – Pipe that was laid first (ID 8076) VS pipe that was laid last (ID 5067):

RING CONDITION ## **Pipe 8076** / ## RING CONDITION ## **Pipe 5067**

Year	Pr. F.	Pr. Leakage	Ave. Leakage. Rate	Year	Pr. F.	Pr. Leakage	Ave. Leakage. Rate
1935	0.000	0.000	0.000	1950	0.000	0.000	0.000
...	...	....	...	...	...	...	...
1950	0.013	0.013	5.507	1966	0.010	0.010	5.507
1951	0.058	0.058	5.507	1967	0.028	0.028	5.507
1952	0.085	0.085	5.507	1968	0.180	0.180	3.672
1953	0.198	0.198	3.672	1969	0.345	0.345	3.672
1954	0.370	0.370	3.672	1970	0.400	0.400	3.672
1955	0.440	0.440	3.672	1971	0.400	0.400	0.000
1956	0.440	0.440	0.000	...	...	...	...
...	...	...	...	2028	0.400	0.400	0.000
2028	0.440	0.440	0.000				

Result: HYPOTHESIS VERIFIED

Test Scenario G4 – Pipe characterized by the highest operating pressure (ID 5067) VS pipe characterised by the lowest operating pressure (ID 21594):

## RING CONDITION ## **Pipe 5067**				## RING CONDITION ## **Pipe 21594**			
Year	**Pr. F.**	**Pr. Leakage**	**Ave. Leakage. Rate**	**Year**	**Pr. F.**	**Pr. Leakage**	**Ave. Leakage. Rate**
1950	0.000	0.000	0.000	1950	0.000	0.000	0.000
...	...	...	...	...	...	...	...
1966	0.010	0.010	5.507	1965	0.010	0.010	4.497
1967	0.028	0.028	5.507	1966	0.033	0.033	4.497
1968	0.180	0.180	3.672	1967	0.040	0.040	4.497
1969	0.345	0.345	3.672	1968	0.130	0.130	2.998
1970	0.400	0.400	3.672	1969	0.288	0.288	2.998
1971	0.400	0.400	0.000	1970	0.378	0.378	2.998
...	...	...	...	1971	0.378	0.378	0.000
2028	0.400	0.400	0.000	...	...	...	...
				2028	0.378	0.378	0.000

Result: HYPOTHESIS VERIFIED

Test Scenario G5 – Pipe characterized by the thinnest cover (ID 21224) VS pipe characterised by the thickest cover (ID 21594):

## RING CONDITION ## **Pipe 21224**				## RING CONDITION ## **Pipe 21594**			
Year	**Pr. F.**	**Pr. Leakage**	**Ave. Leakage. Rate**	**Year**	**Pr. F.**	**Pr. Leakage**	**Ave. Leakage. Rate**
1950	0.000	0.000	0.000	1950	0.000	0.000	0.000
1965	0.008	0.008	2.998	...	...	...	...
1966	0.030	0.030	2.998	1965	0.010	0.010	4.497
1967	0.030	0.030	0.000	1966	0.033	0.033	4.497
...	...	...	...	1967	0.040	0.040	4.497
2028	0.030	0.030	0.000	1968	0.130	0.130	2.998
				1969	0.288	0.288	2.998
				1970	0.378	0.378	2.998
				1971	0.378	0.378	0.000
				...	...	...	...
				2028	0.378	0.378	0.000

Result

Here the interpretation is less clear; if the software considered the equilibrium of the soil mass above the pipe with development of tangential stress along failure surfaces, pipes having thicker covers would be subjected to lower stresses; on the contrary, if this factor were not taken into account (which is very likely), pipes having thicker covers would be subjected to higher stresses. In the case under survey, considering the very thin cover (even if the first hypothesis were true) surfaces would be outcropping and the pipe would only be subjected to the weight of soil itself (the selected calculation mode does not contemplate road loads). Therefore, we can expect that thinner covers (21224) correspond to lower loads and lower failure probabilities; nevertheless the gap between the two results seems excessive.

In conclusion, the software appears to provide results meeting expectations, and the methodology can be considered as applicable to calculate failure risk probabilities.

AQUASERV, ROMANIA

Software application

In order to evaluate the developed WATERPIPE DSS software for usefulness, user friendliness, ease of installation, trustworthiness, some pipes included in the Aquaserv water distribution network in Tirgu Mures in Romania have been selected and introduced to the DSS software. The pipes selected are presented in Table 8.2.

Analysis involved different type of pipes, but grouped for different testing scenarios in same categories. Same as for other similar software applications, WATERPIPE DSS requires detailed information regarding the water distribution network including pipe materials, environmental conditions, water characteristics, road loading conditions, soil properties. Due to the lack of a number of information needed for calculations some default values obtained from the literature was used.

It is fact that existing GIS information are not 100% accurate. Our GIS experts consider an approx. 70% accuracy of the existing GIS data.

Table 8.2 Pipes selected to evaluate the developed WATERPIPE DSS – Tirgu Mures.

Network Id. No.	Pipe Material	Year of Construction	External Diameter (mm)	Initial Wall Thickness (mm)	Depth of Cover (m)	Length of Pipe (m)	Surge Pressure (kN/m^2)	Type of Joint	Working Pressure (kN/m^2)	Water Table Average Depth (m)	Backfill Type	Type of Water	Street	Pipe data used in
163	pe	2006	110	10	2.4	261.06	600	continous	365	3.8	sandy clay	upland	tudor vladimirescu	Test Scenario 1
418	steel	1965	114	4.25	1.5	194.79	600	bell	435	2.8	black clay	upland	parmatei	Test Scenario 1
750	steel	1970	660	8.8	1.7	459.45	600	flexible	520	3.2	gravel with sand	upland	n. balcescu	Test Scenario 2
589	steel	1965	660	8.8	3	366.97	600	flexible	520	4.2	grey silty clay	upland	kossuth lajos	Test Scenario 2
161	steel	1965	114	4.25	1.4	148.77	600	flexible	420	3.8	sandy clay	upland	bobalna	Test Scenario 3
284	steel	1965	114	4.25	1.5	170.58	600	flexible	446	2.5	silty sand	upland	lebedei	Test Scenario 3
870	cast iron	1965	144	4.8	1.7	268.08	600	bell	410	4	argillaceous and sand	upland	argesului	Test Scenario 4
866	steel	1961	140	4.5	1.3	112.84	600	flexible	388	4	argillaceous and sand	upland	trebely	Test Scenario 4
284	steel	1965	114	4.25	1.5	170.58	600	flexible	446	2.5	silty sand	upland	lebedei	Test Scenario 5
427	steel	1970	114	4.25	1.0	218.21	600	flexible	449	3.2	gravel with sand	upland	n. balcescu	Test Scenario 5
161	steel	1965	114	4.25	1.4	148.77	600	flexible	420	3.8	sandy clay	upland	bobalna	Test Scenario 5
656	steel	1985	114	4.25	1.2	273.29	600	flexible	534	1.5	sandy clay	upland	b-dul 1 decembrie 1918	Test Scenario 5

The following testing scenarios were selected:

Test Scenario T1: pipe "in worst condition" vs. pipe "in best condition"
Test Scenario T2: pipe that was laid first vs. pipe that was laid last
Test Scenario T3: pipe characterized by the thinnest cover vs. pipe characterized by the thickest cover
Test Scenario T4: pipe having the same diameter but was laid in different years
Test Scenario T5: pipe having the same initial thickness

The detailed results of the test scenarios are presented below.

Verification tests

Test Scenario T1-Pipe "in worst condition" (ID 418) vs pipe "in best condition" (ID 163)
RING CONDITION ## **Pipe "418"** (Name P.Armatei-Dext = 114 mm, steel pipe)

Year	Pr. F.	Pr. Leakage	Ave. Leakage. Rate	Year	Pr. F.	Pr. Leakage	Ave. Leakage. Rate
1965	0.000	0.000	0.000	2011	0.720	0.633	4.771
...	...	...	...	2012	0.785	0.698	4.316
1999	0.023	0.023	3.620	2013	0.883	0.795	5.429
2000	0.023	0.023	0.000	2014	0.883	0.795	0.000
2001	0.055	0.055	5.429	2015	0.908	0.820	5.067
2002	0.110	0.110	5.429	2016	0.943	0.855	5.429
2003	0.110	0.110	0.000	2017	0.943	0.855	0.000
2004	0.130	0.110	0.000	2018	0.943	0.855	0.000
2005	0.180	0.150	5.429	2019	0.975	0.888	5.429
2006	0.333	0.295	4.556	2020	0.975	0.888	0.000
2007	0.373	0.335	3.846	2021	0.975	0.888	0.000
2008	0.455	0.388	5.429	2022	0.975	0.888	0.000
2009	0.548	0.460	4.556	2023	1.000	0.913	5.429
2010	0.610	0.523	5.140				

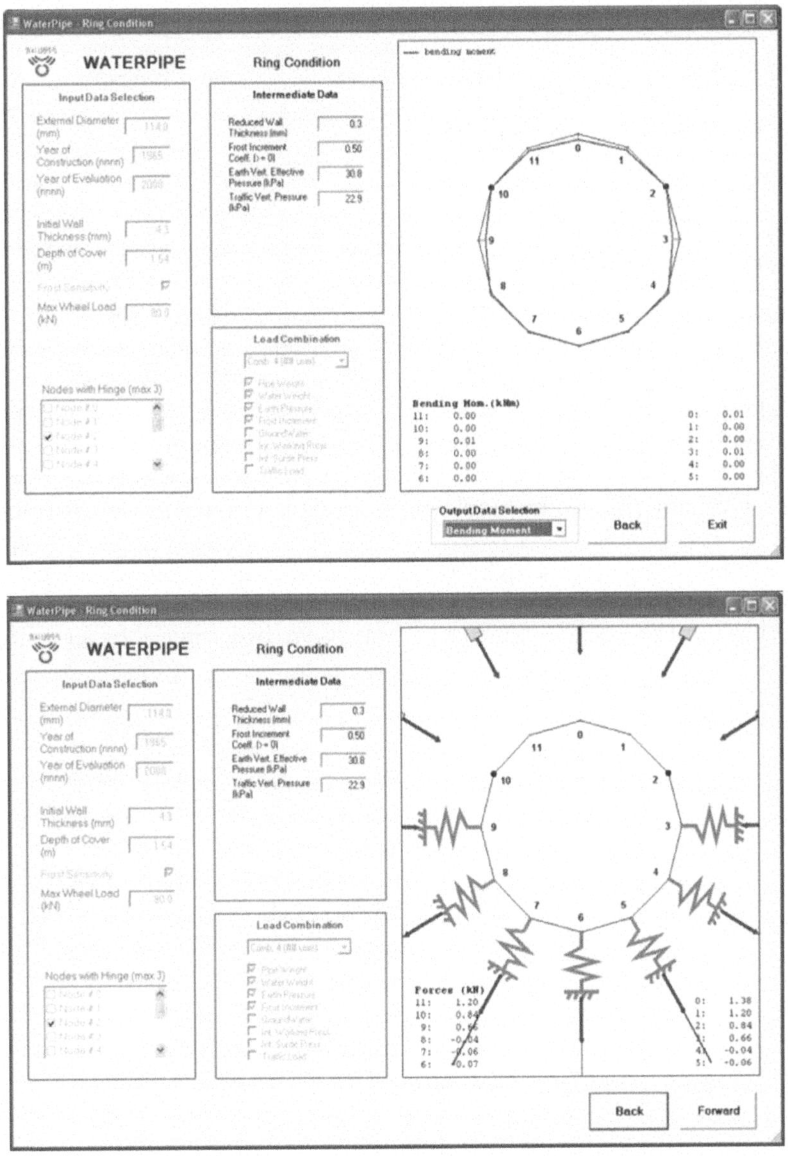

Figure 8.10 Life Cycle Calculation Results for "Pipe 418".

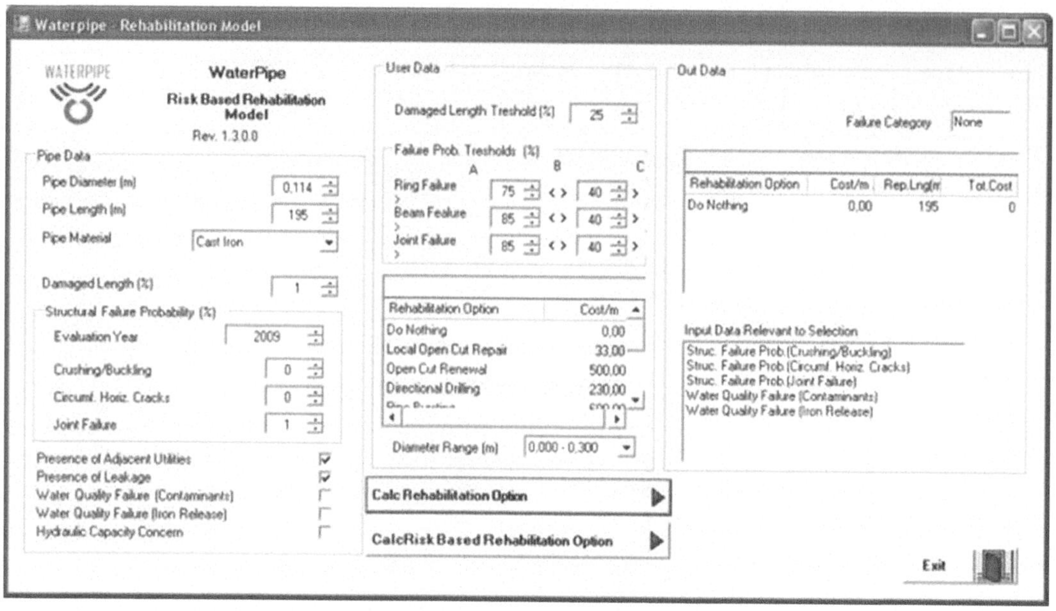

Figure 8.11 Rehabilitation Module results.

RING CONDITION ## **Pipe 163** (Name T.Vladimirescu-Dext = 110 mm, PE pipe)

Year	Pr. F.	Pr. Leakage	Ave. Leakage. Rate
2006	0.000	0.000	0.000
...	...	...	
2028	0.000	0.000	0.000
2029	0.000	0.000	0.000

Expected results: Pipe "in worst condition" fails earlier than pipe in "best condition".

DSS testing results: Hypothesis confirmed that "Pipe 418" probability of failure happens earlier than probability of failure for "Pipe 163".

The Rehabilitation module for Pipe 418 in the presence of a joint failure, adjacent utilities present, presence of leakage, with damaged length from 0 to 99% suggested to "DO nothing". It was observed that only Water quality issues will conduct to a method for rehabilitation (Directional Drilling in this case) with the default failure probability threshold values. However, the expected option (Local Repair), used for repairing the detected failure was not suggested in any conditions.

Field results: All the testing equipments (correlator, e-spade radar, earth microfons) detected clearly a leaking pipe. After digging out, a joint failure was found as it is shown in Figure 8.13. The splashing water flowed underground directly to the sewage network. The position of the failure was accuretly detemined by all the equipments.

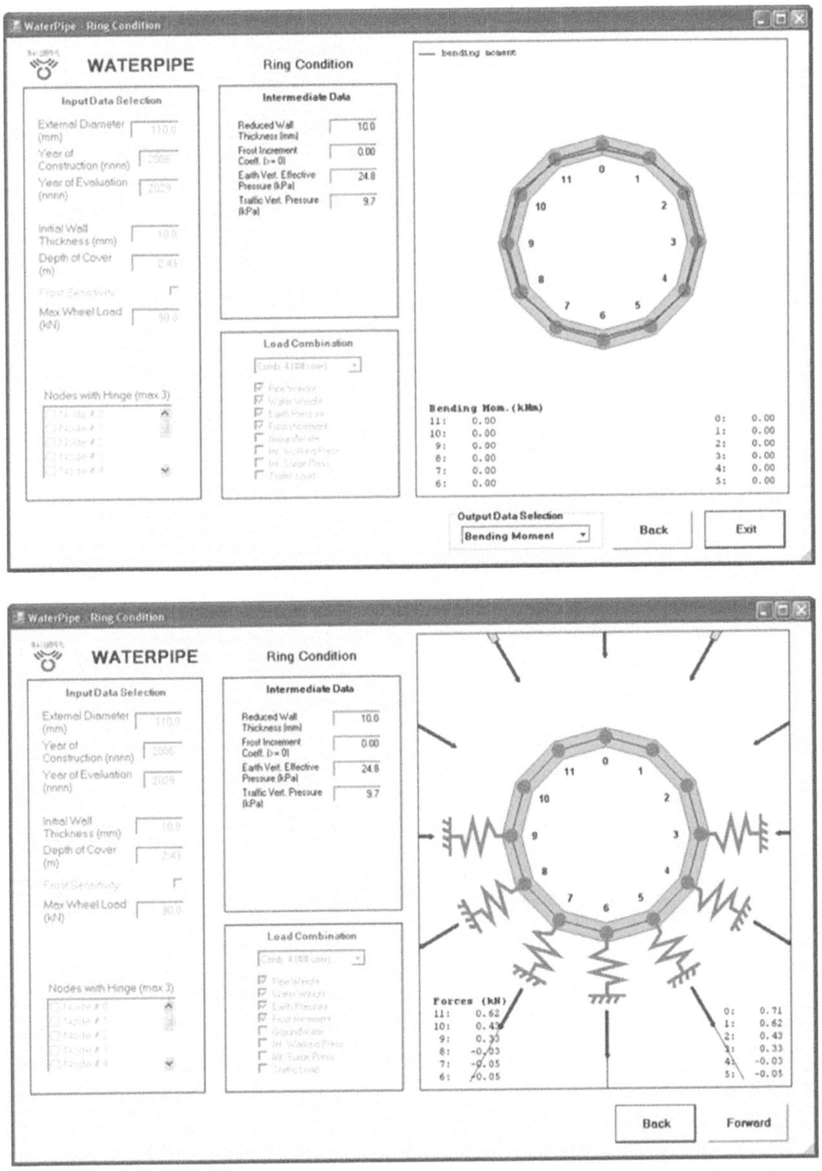

Figure 8.12 Life Cycle Calculation Results for "Pipe 163".

Figure 8.13 Photo of the joint failure detected on the Pipe 418.

Test scenario T2-Pipe that was laid first (ID 589) VS pipe that was laid last (ID 750)

Pipe 750 Name N.Balcescu-Dext = 660 mm, steel pipe ## RING CONDITION ##				Pipe 589 Name Kossuth Lajos-Dext = 660 mm, steel pipe ## RING CONDITION ##			
Year	**Pr. F.**	**Pr. Leakage**	**Ave. Leakage. Rate**	**Year**	**Pr. F.**	**Pr. Leakage**	**Ave. Leakage. Rate**
1970	0.000	0.000	0.000	1965	0.000	0.000	0.000
1971	0.000	0.000	0.000	1966	0.000	0.000	0.000
...	...	...	....	...	...	...	....
2029	0.000	0.000	0.000	2029	0.000	0.000	0.000

Expected results: Pipe "laid first" fails earlier than pipe in "laid last". Ring condition for Pipe "589" starting at least from year 2004, as it is showed on the GIS records – see Figure 8.14 – (several recorded repairs starting from the year 2004).

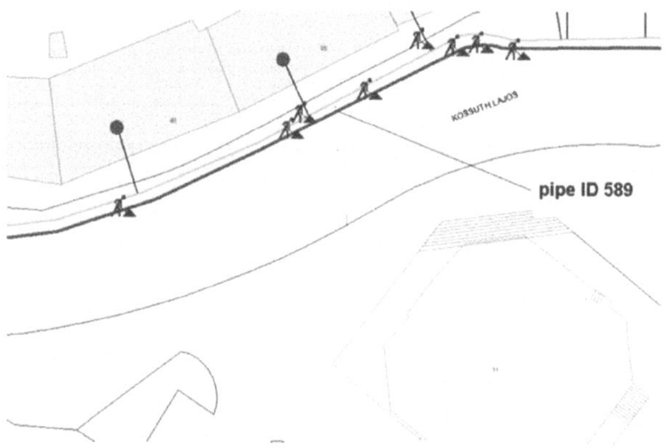

Figure 8.14 GIS map showing pipe id. 589.

DSS testing result: The same results were produced for both pipes: both pipes will not fail up to year 2029. We consider the calculations are not correct due to the fact that an error message reported a not recognized pipe material (steel). However, GIS map showing that on the pipe id 589 were several repairs due to the fact that the pipe follows the street curve, and the pipe was constructed from several small segments, which cannot be entered into DSS.

Field results: All the testing equipments (correlator, e-spade radar, earth microfons) detected a several possible leaks. After digging out, several joint failured were found (please see Figure 8.15). The splashing water flowed underground directly to the sewage network. The position of the failures was accuretly detemined by all the equipments.

Figure 8.15: Photos of the joint failures found on the Pipe 589.

Test scenario T3-Pipe characterized by the thinnest cover (ID 161) VS pipe characterized by the thickest cover (ID 284)
RING CONDITION ## **Pipe 161** (Name Bobalna-Dext = 114 mm, steel pipe, 1.4 m)

Year	Pr. F.	Pr. Leakage	Ave. Leakage. Rate	Year	Pr. F.	Pr. Leakage	Ave. Leakage. Rate
1965	0.000	0.000	0.000	2013	0.655	0.388	3.997
...	...	...	...	2014	0.750	0.455	4.543
2003	0.055	0.025	4.543	2015	0.878	0.508	4.326
2004	0.073	0.025	0.000	2016	0.878	0.508	0.000
2005	0.073	0.025	0.000	2017	0.908	0.525	3.028
2006	0.073	0.025	0.000	2018	0.945	0.530	4.543
2007	0.108	0.053	4.405	2019	0.958	0.543	4.543
2008	0.178	0.100	3.267	2020	0.970	0.543	0.000
2009	0.235	0.153	4.326	2021	0.990	0.543	0.000
2010	0.310	0.215	3.997	2022	0.990	0.543	0.000
2011	0.413	0.238	4.543	2023	0.990	0.543	0.000
2012	0.518	0.325	3.504	2024	0.990	0.543	0.000
				2025	1.000	0.543	0.000

RING CONDITION Pipe 284 (Name Lebedei -Dext = 114 mm, steel pipe, 1.5 m)

Year	Pr. F.	Pr. Leakage	Ave. Leakage. Rate	Year	Pr. F.	Pr. Leakage	Ave. Leakage. Rate
1965	0.000	0.000	0.000	2011	0.828	0.788	4.539
....	....	....	....	2012	0.878	0.838	4.213
1999	0.003	0.003	3.121	2013	0.925	0.885	4.270
2000	0.003	0.003	0.000	2014	0.925	0.885	0.000
2001	0.235	0.235	4.681	2015	0.945	0.905	4.096
2002	0.255	0.255	4.681	2016	0.968	0.928	3.988
2003	0.255	0.255	0.000	2017	0.968	0.928	0.000
2004	0.268	0.255	0.000	2018	0.973	0.933	4.681
2005	0.303	0.290	4.681	2019	0.990	0.950	4.681
2006	0.585	0.565	3.291	2020	0.990	0.950	0.000
2007	0.613	0.580	4.421	2021	0.990	0.950	0.000
2008	0.645	0.613	4.201	2022	0.990	0.950	0.000
2009	0.710	0.678	4.261	2023	1.000	0.960	4.681
2010	0.768	0.733	4.397				

Expected results: Pipes having thicker covers would fail later.

DSS testing result: The results produced were not as we expected. The calculations are probably not correct due to the fact that an error message (not recognized pipe material (steel pipe)) was reported by the DSS . However, we should also mention that the GIS data regarding year of construction are not fully accurate.

Test scenario T4-Pipe having the same diameter(ID 870;ID 866) but that was laid in different years

RING CONDITION ## Pipe 870 (Name Argesului-Dext = 144 mm, cast iron)

Year	Pr. F.	Pr. Leakage	Ave. Leakage. Rate	Year	Pr. F.	Pr. Leakage	Ave. Leakage. Rate
1965	0.000	0.000	0.000	1980	0.915	0.258	0.000
...	...	...	...	1981	0.923	0.263	5.645
1976	0.265	0.108	5.645	1982	0.998	0.338	5.645
1977	0.475	0.128	5.645	1983	0.998	0.338	0.000
1978	0.770	0.255	5.645	1984	1.000	0.340	3.763
1979	0.780	0.258	5.645				

RING CONDITION ## **Pipe 866** (Name Trebely-Dext = 140 mm, steel pipe)

Year	Pr. F.	Pr. Leakage	Ave. Leakage. Rate	Year	Pr. F.	Pr. Leakage	Ave. Leakage. Rate
1961	0.000	0.000	0.000	2013	0.793	0.650	5.060
...	...	...	...	2014	0.798	0.650	0.000
2006	0.103	0.103	4.225	2015	0.798	0.650	0.000
2007	0.148	0.148	3.571	2016	0.798	0.650	0.000
2008	0.230	0.230	5.249	2017	0.803	0.650	0.000
2009	0.330	0.315	5.357	2018	0.820	0.668	3.571
2010	0.448	0.430	3.882	2019	0.820	0.668	0.000
2011	0.655	0.635	4.573	2020	0.820	0.668	0.000
2012	0.765	0.635	0.000	2021	1.000	0.743	3.571

Expected results: Pipe laid first fails earlier.

DSS testing result: The results produced were as we expected.

Field test results: Night time measurements indicate a possible leakage on pipe 866. All the equipments and teams detected two (2) possible leaks in a distance of 3.5 m between them. A valve on the pipe was found which was not showed in the GIS, as shown in Figure 8.16.

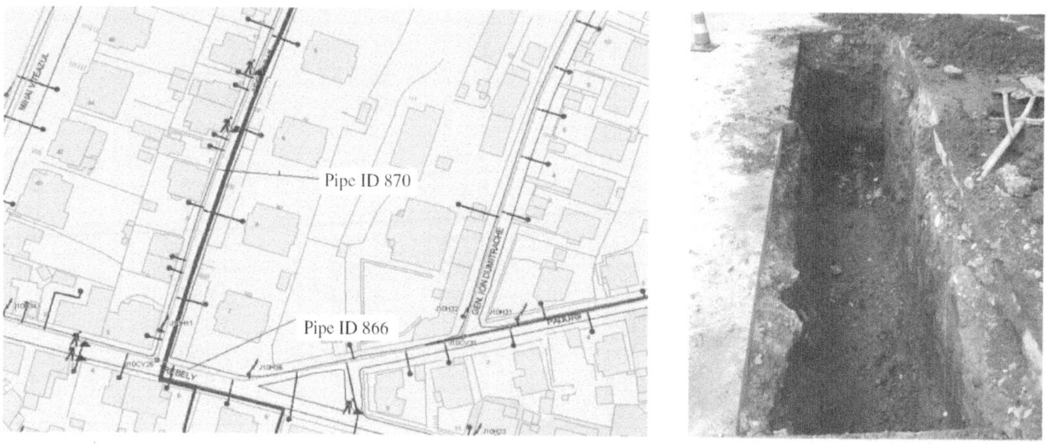

Figure 8.16 (Left) GIS map showing pipes id. 870 and id 866(Right) Hidden valve found on the pipe id 866.

Test scenario T5-Pipe having the same initial thickness (ID 284;ID 427;ID 161;ID 656). The GIS maps for the pipes with ID 284, 427, 161 and 656 are shown in the Figures 8.17–8.19.

RING CONDITION ## **Pipe 284** (Name Lebedei-Dext = 114 mm)

Year	Pr. F.	Pr. Leakage	Ave. Leakage. Rate	Year	Pr. F.	Pr. Leakage	Ave. Leakage. Rate
1965	0.000	0.000	0.000	2011	0.828	0.788	4.539
...	...	...	...	2012	0 .878	0.838	4.213
1999	0.003	0.003	3.121	2013	0.925	0.885	4.270
2000	0.003	0.003	0.000	2014	0.925	0.885	0.000
2001	0.235	0.235	4.681	2015	0.945	0.905	4.096
2002	0.255	0.255	4.681	2016	0.968	0.928	3.988
2003	0.255	0.255	0.000	2017	0.968	0.928	0.000
2004	0.268	0.255	0.000	2018	0.973	0.933	4.681
2005	0.303	0.290	4.681	2019	0.990	0.950	4.681
2006	0.585	0.565	3.291	2020	0.990	0.950	0.000
2007	0.613	0.580	4.421	2021	0.990	0.950	0.000
2008	0.645	0.613	4.201	2022	0.990	0.950	0.000
2009	0.710	0.678	4.261	2023	1.000	0.960	4.681
2010	0.768	0.733	4.397				

Figure 8.17 GIS map showing pipes id. 284.

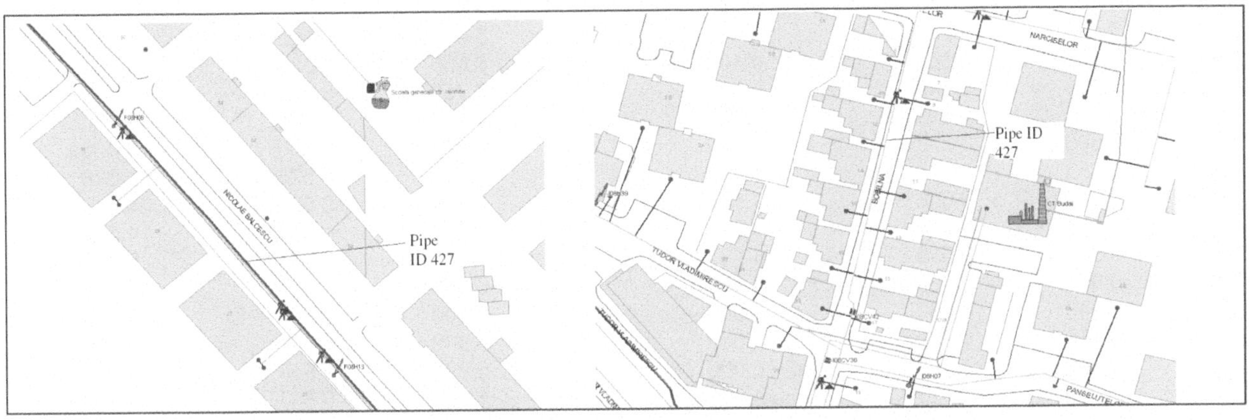

Figure 8.18 (Left) GIS map showing pipes id. 427, (Right) GIS map showing pipes id. 161.

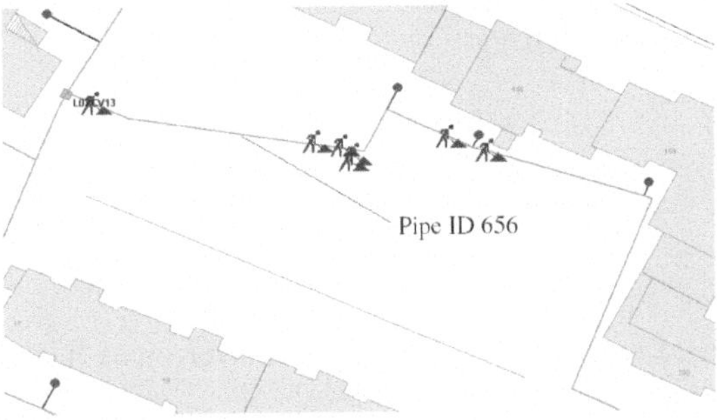

Figure 8.19 GIS map showing pipes id. 656.

RING CONDITION
Pipe 427 (Name N.Balcescu-Dext = 114 mm)

Year	Pr. F.	Pr. Leakage	Ave. Leakage. Rate
1970	0.000	0.000	0.000
...	...	...	....
2005	0.003	0.003	3.131
2006	0.003	0.003	0.000
2007	0.003	0.003	0.000
2008	0.005	0.005	4.697
2009	0.005	0.005	0.000
2010	0.020	0.005	0.000
2011	0.068	0.050	4.697
2012	0.128	0.055	4.697
2013	0.188	0.093	4.279
2014	0.273	0.153	3.653
2015	0.375	0.210	3.540
2016	0.435	0.253	4.605
2017	0.605	0.365	4.105
2018	0.725	0.448	4.697
2019	0.800	0.523	4.384
2020	0.800	0.523	0.000
2021	0.853	0.565	4.144
2022	0.915	0.628	4.509
2023	0.915	0.628	0.000
2024	0.918	0.630	4.697
2025	0.948	0.660	4.697
2026	0.948	0.660	0.000
2027	0.948	0.660	0.000
2028	0.948	0.660	0.000
2029	1.000	0.713	4.697

Pipe 161 (Name Bobalna-Dext = 114 mm)

Year	Pr. F.	Pr. Leakage	Ave. Leakage. Rate
1965	0.000	0.000	0.000
...	...	...	....
2001	0.030	0.000	0.000
2002	0.030	0.000	0.000
2003	0.055	0.025	4.543
2004	0.073	0.025	0.000
2005	0.073	0.025	0.000
2006	0.073	0.025	0.000
2007	0.108	0.053	4.405
2008	0.178	0.100	3.267
2009	0.235	0.153	4.326
2010	0.310	0.215	3.997
2011	0.413	0.238	4.543
2012	0.518	0.325	3.504
2013	0.655	0.388	3.997
2014	0.750	0.455	4.543
2015	0.878	0.508	4.326
2016	0.878	0.508	0.000
2017	0.908	0.525	3.028
2018	0.945	0.530	4.543
2019	0.958	0.543	4.543
2020	0.970	0.543	0.000
2021	0.990	0.543	0.000
2022	0.990	0.543	0.000
2023	0.990	0.543	0.000
2024	0.990	0.543	0.000
2025	1.000	0.543	0.000

RING CONDITION ## **Pipe 656** (Name 1 Dec. 1918-Dext = 114 mm)

Year	Pr. F.	Pr. Leakage	Ave. Leakage. Rate
1985	0.000	0.000	0.000
...	...	...	...
2021	0.010	0.010	3.415
2022	0.018	0.010	0.000
2023	0.018	0.010	0.000
2024	0.025	0.018	5.122
2025	0.025	0.018	0.000
2026	0.055	0.018	0.000
2027	0.108	0.045	5.122
2028	0.228	0.065	5.122
2029	0.288	0.070	5.122

Expected results: Comparing field results and GIS history data for pipes having same initial thickness and same material.

DSS testing result: The results obtained are the followings:

(1) For pipe id 284 Lebedei GIS shows a repair work in year 2004, and the DSS estimates that the first failure starting from 1999. We consider the result is OK.

(2) For pipe id 427 Balcescu GIS shows repair works in year 2006 and 2008 and the DSS estimates the first failure starting from 2005. We consider the result is OK.

(3) For pipe id 161 Bobalna GIS shows repair work in year 2007 and the DSS estimates the first failure starting from 2001. We consider the result is OK.

(4) For pipe id 656 1918 Decembrie GIS shows repair works starting from 2004 and the DSS estimates the first failure starting from 2021. The result is not OK. However it is also possibile that the layed pipe material was not respecting the standards.

RISA, GREECE

Input information

Data was collected by EYDAP, the Water and Sewerage Company of Athens, Greece, relating to the physical characteristics as well as the break history for a 6 year period, 2003–2009, for a 2.385 km long, 900 mm diameter steel trunk main currently in operation in EYDAP's bulk supply network, Figure 8.20. The pipeline was constructed in 1962 with an average depth of 1.2 m below the ground surface and it is operating with a maximum internal pressure of 10 bar. The pipe is of welded construction with no cathodic protection provided. Trafic wheel loading is taken as not being more than 100 kN. The data were given to RISA and were entered into the DSS. The model was run and some alterations to the build in values of the DSS with regard to materials were made based on the user's recommendations as well as adding default parameters for steel pipelines.

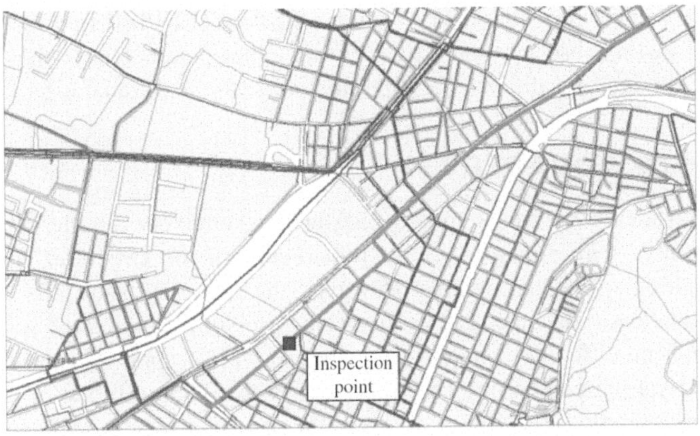

Figure 8.20 Transmission pipeline in EYDAP – Athens-Greece.

Verification tests

The results from the DSS showed that the pipeline will not fail up to year 2029. Even by decreasing the pipe wall thickness down to 8 mm, still no failure occurred. Some probable explanations are provided below:

- No surges due to the stable operation of the pipe (almost constant pressure – daily variation from 90 to 100 m)
- High quality of materials and construction works
- In general, non – corrosive soils in the Athens area and treated water of high quality (pH of 7,6 non aggressive)
- Current thickness of pipe measured downstream from leakage events. Therefore, more appropriate to run the DSS for shorter lengths of this particular pipe – at least 4, after having measured the current thickness of each one in case of differentiation of corrosion parameters.
- Further investigation and verification of level of leakage and condition of pipe by using advanced leakage detection technologies (hydrophones).

The pipeline has in fact failed locally due to pit corrosion and the points of failure are shown on Figure 8.21.

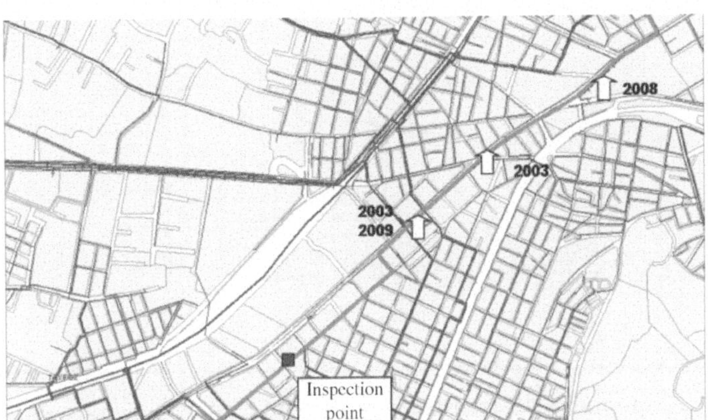

Figure 8.21 Transmission pipeline in EYDAP- Points of local failure.

CONCLUSIONS

Conclusions reached by AMGA

The conclusions drawn by the users of AMGA can be summarised to the following:

(a) The software, as a whole, can be considered as simple, user-friendly and useful. It provides the user with valuable data to plan rehabilitation actions, combining hydraulic information (such as failure probabilities, leakage level and presence of pollutants) and structural information (such as level of stresses, values of the stress characteristics and of safety factors).

(b) Like other similar software applications, WATERPIPE requires a detailed knowledge of the network, the materials, the water characteristics, the road loads, the soil properties, and characterises each element with the related coefficients describing its behaviour. It is reasonable to assume that it could be successfully applied, after a short period needed by the user to fully understand the software characteristics and functions.

(c) Define a methodology to help decision makers to use DSS WATERPIPE in a more effective way avoiding to analyse each pipe but just a subset of more critical ones. Other parameters could be evaluated for the purpose such as the pipe age, the load conditions, the pipe thickness, the leak reparation frequency analysis.

Conclusions reached by AQUASERV

The software produced is considered as an integrated package, is relatively simple, user-friendly and useful. The produced results provide the user with valuable information regarding the structural condition of the network and support for planning rehabilitation works and options. However, the following shortfalls were identified:

(a) The DSS software can be easily installed but we find some problems running the software under Microsoft Vista (data transfer problem from different modules) and some old P4 PCs under Microsoft XP. We suggest specifying the minimum system requirements in order to run properly the DSS software.

(b) The DSS software does not allow multiple accesses to the database. Due to this situation working with the DSS and the database behind is mainly limited to one computer/one user.

(c) Using the software is relatively easy. However the lack of an online tutorial to use the software, the lack of context help menu, and the incomplete translation from German into English language of several context menus causes problems using the software (Figure 8.22).

(d) The average Leakage rate calculations seem not to be correct, due to the fact that the average leakage rate results plotted on the time axis shows an "unnatural evolution" of the leakage rate in time (Figure 8.23).

(e) Pipe materials in different modules are not fully correlated. In the Structural reliability module for e.g. steel pipes cannot be defined (Figure 8.24).

Due to this situation the module gives the following error messages (Figure 8.25):

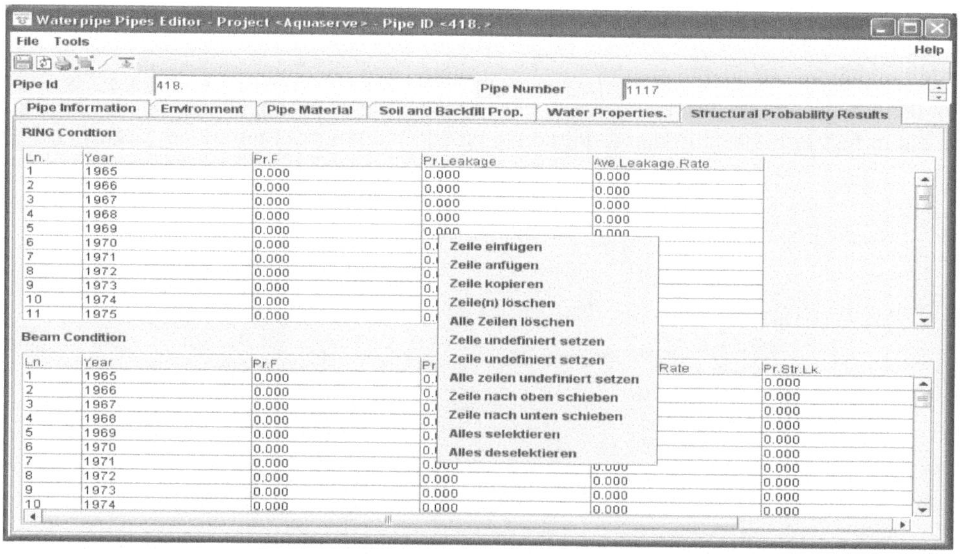

Figure 8.22 Translation problem in a context menu.

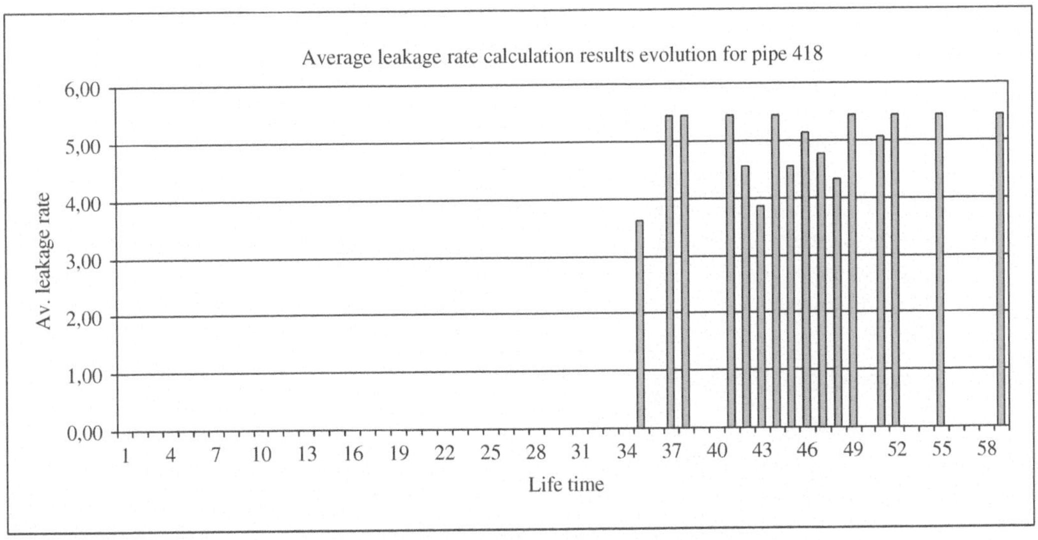

Figure 8.23 Graphical evolution of the Average leakage rate for pipe.

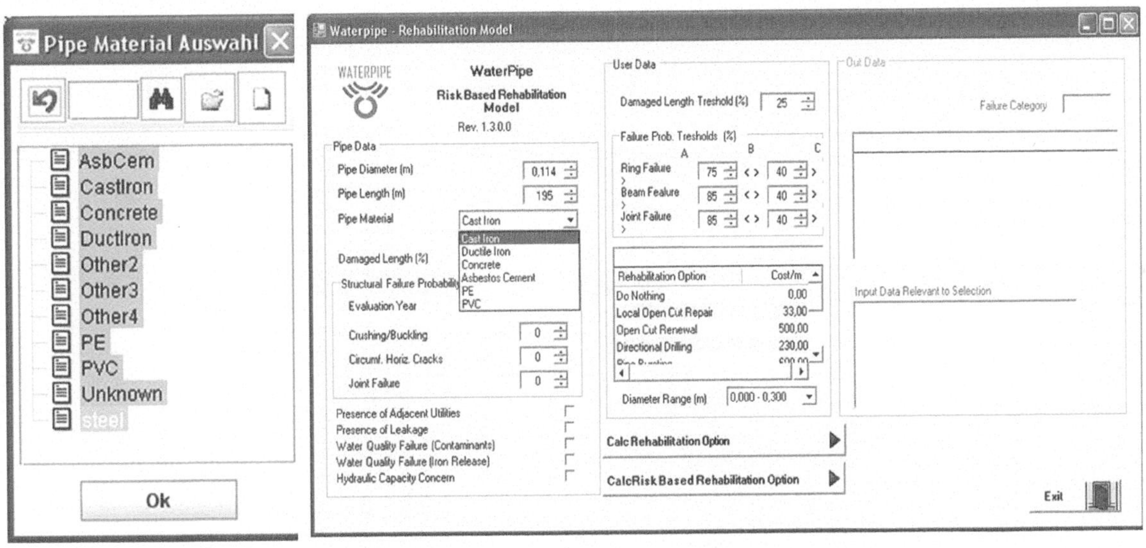

Figure 8.24 Pipe materials option in different modules.

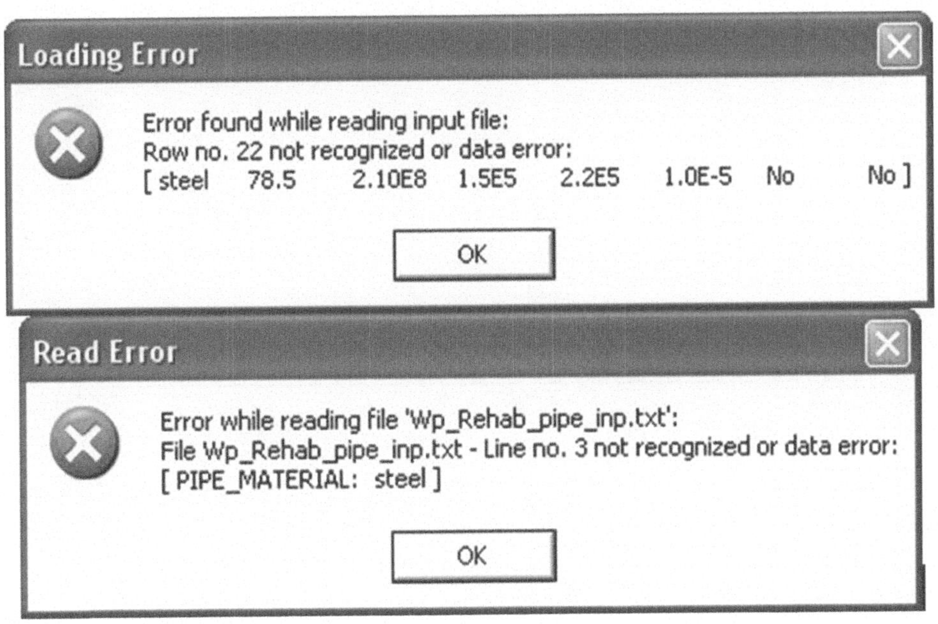

Figure 8.25 Error messages due to Pipe materials correlation problem.

After confirming the error message, the system seems to work normally, however for steel pipes the calculation results are probably erroneous.

(a) The Risk based Rehabilitation option calculation produced no results.
(b) Default values for several coefficients used are different from the coefficients presented in the methodologies e.g. coefficients for pipe materials.
(c) The DSS currently is configured to perform, analysis on a single pipe.
(d) The DSS currently does not have a method to indicate critical pipes in the network.
(e) In the water distribution network there are several pipes which were repaired in the past. For example 70 m of a 100 m total length pipe was replaced. As a result the probability of failure for that pipe was decreased. At the moment the DSS does not permit to specify this situation and all the calculations are done with the initial pipe condition.

Conclusions reached by EYDAP

The following conclusions have been reached by EYDAP with regard to the use of the DSS for transmission mains:

- Results from the DSS would be closer to real conditions for short lengths of pipeline.
- It is extremely important to have accurate records to enter in to the DSS.
- Maintain accurate historical results of leaks / bursts and obtain information on the actual condition of the pipe.

IWA Publishing's authorised EU representative for General Product Safety Regulations is Diane D'Arras, 15 rue Duret, 75116 Paris, France, e-mail: safety@iwap.co.uk.

Printed and bound by CPI Group (UK) Ltd, Croydon, CR0 4YY
12/05/2026
02108874-0017